Marius Schollmeier

Optimization and Control of Laser-Accelerated Proton Beams

Marius Schollmeier

Optimization and Control of Laser-Accelerated Proton Beams

Optimierung und Kontrolle laserbeschleunigter Protonenstrahlen

Südwestdeutscher Verlag für Hochschulschriften

Impressum/Imprint (nur für Deutschland/only for Germany)
Bibliografische Information der Deutschen Nationalbibliothek: Die Deutsche Nationalbibliothek verzeichnet diese Publikation in der Deutschen Nationalbibliografie; detaillierte bibliografische Daten sind im Internet über http://dnb.d-nb.de abrufbar.
Alle in diesem Buch genannten Marken und Produktnamen unterliegen warenzeichen-, marken- oder patentrechtlichem Schutz bzw. sind Warenzeichen oder eingetragene Warenzeichen der jeweiligen Inhaber. Die Wiedergabe von Marken, Produktnamen, Gebrauchsnamen, Handelsnamen, Warenbezeichnungen u.s.w. in diesem Werk berechtigt auch ohne besondere Kennzeichnung nicht zu der Annahme, dass solche Namen im Sinne der Warenzeichen- und Markenschutzgesetzgebung als frei zu betrachten wären und daher von jedermann benutzt werden dürften.

Verlag: Südwestdeutscher Verlag für Hochschulschriften GmbH & Co. KG
Dudweiler Landstr. 99, 66123 Saarbrücken, Deutschland
Telefon +49 681 37 20 271-1, Telefax +49 681 37 20 271-0
Email: info@svh-verlag.de

Approved by: Darmstadt, TU, Diss., 2008

Herstellung in Deutschland:
Schaltungsdienst Lange o.H.G., Berlin
Books on Demand GmbH, Norderstedt
Reha GmbH, Saarbrücken
Amazon Distribution GmbH, Leipzig
ISBN: 978-3-8381-0857-5

Imprint (only for USA, GB)
Bibliographic information published by the Deutsche Nationalbibliothek: The Deutsche Nationalbibliothek lists this publication in the Deutsche Nationalbibliografie; detailed bibliographic data are available in the Internet at http://dnb.d-nb.de.
Any brand names and product names mentioned in this book are subject to trademark, brand or patent protection and are trademarks or registered trademarks of their respective holders. The use of brand names, product names, common names, trade names, product descriptions etc. even without a particular marking in this works is in no way to be construed to mean that such names may be regarded as unrestricted in respect of trademark and brand protection legislation and could thus be used by anyone.

Publisher: Südwestdeutscher Verlag für Hochschulschriften GmbH & Co. KG
Dudweiler Landstr. 99, 66123 Saarbrücken, Germany
Phone +49 681 37 20 271-1, Fax +49 681 37 20 271-0
Email: info@svh-verlag.de

Printed in the U.S.A.
Printed in the U.K. by (see last page)
ISBN: 978-3-8381-0857-5

Copyright © 2009 by the author and Südwestdeutscher Verlag für Hochschulschriften GmbH & Co. KG and licensors
All rights reserved. Saarbrücken 2009

Zusammenfassung

Die Bestrahlung von mikrometerdünnen Metallfolien durch moderne Hochenergiekurzpulslaser mit Intensitäten größer als $10^{18}\,\text{W/cm}^2$ führt unter anderem zur Beschleunigung von Ionenstrahlen mit Energien im Bereich von Megaelektronenvolt (MeV). Der Laserpuls beschleunigt zuerst Elektronen auf relativistische Energien, die dann durch die Folie propagieren. Sobald die Elektronen die Folie auf der Rückseite verlassen, wird ein elektrisches Feld mit einer Feldstärke um $10^{12}\,\text{V/m}$ generiert. Adsorbierte Protonen von der Folienoberfläche können somit sehr effektiv in Richtung der Targetnormalen beschleunigt werden. Diese derart generierten quasineutralen Strahlen bestehen aus mehr als 10^{12} Protonen in einem kurzen, ca. Pikosekunden andauernden Puls. Mögliche Anwendungsgebiete sind die Diagnostik dichter Plasmen, die Nutzung als kompakter Injektor für Teilchenbeschleuniger, das Anregen kernphysikalischer Prozesse, die Energiegewinnung durch schnelle Zündung bei der Trägheitsfusion sowie ein eventueller Einsatz in der Krebstherapie mit Ionenstrahlen.

So verfügen laserbeschleunigte Ionenstrahlen über einige Strahleigenschaften die weit besser sind als die von Ionenstrahlen aus herkömmlichen Ionenquellen. Dies motiviert den Einsatz als Ionenquelle der nächsten Generation. Bis heute existiert jedoch kein vollständiges Modell der Ionenbeschleunigung mit Lasern welches für Abschätzungen aller Strahlparameter verwendet werden kann. Die Entwicklung von Anwendungen erfordert jedoch eine genaue Kenntnis der Orts- und Impulsverteilung (Phasenraum) der Ionen sowie eine möglichst präzise Modellierung des Beschleunigungsvorgangs. Hierzu wurde die Messtechnik der abbildenden Spektroskopie mit radiochromischen Filmen "Radiochromic Film Imaging Spectroscopy (RIS)" entwickelt. Die für RIS benötigten Dosimetriefilme wurden am Tandembeschleuniger des Max-Planck-Instituts für Kernphysik in Heidelberg für Protonen absolut kalibriert.

RIS verwendet die Methode der Ionenstrahlmanipulation durch mikrometergroße Verformung der Folienoberfläche. Es wurde eine Methode entwickelt, um äquidistante, mikrometergroße Gräben (Abstand entweder 3, 5 oder 10 Mikrometer) in die Oberfläche von dünnen Folien (Dicke 5 bis 50 Mikrometer) einzubringen. Dies geschieht per Ultrahochpräzisions-zerspanung eines Trägermaterials sowie darauf folgender galvanischer Abscheidung der Folien und anschließendem Ätzen des Trägermaterials.

Die mikrostrukturierten Folien wurden in Experimenten am Petawatt High Energy Laser for Heavy Ion eXperiments (PHELIX) des GSI Helmholtzzentrums für Schwerionenforschung GmbH (Darmstadt, März 2006), in zwei Experimentkampagnen am TRIDENT-Laser des Los Alamos National Laboratory (New Mexico, USA, Mai 2005 und April 2006), am 100 TW Laser des Laboratoire pour l'Utilisation des Lasers Intenses (École Polytechnique, Palaiseau, Frankreich, Juni 2006) sowie am Z-Petawatt Laser der Sandia National Laboratories (New Mexico, USA, Dezember 2007) erfolgreich eingesetzt. Die Auswertung der Daten bestätigte nicht nur die Erkenntnisse früherer Experimente, sondern erlaubte zudem Rückschlüsse auf die der Beschleunigung zugrunde liegenden elektrischen Felder.

Die Ergebnisse von RIS flossen in die Entwicklung des Charged Particle Transfer (CPT) Code ein, mit dem die Ionenbeschleunigung von der Rückseite der Folie dreidimensional simuliert werden kann. Mit CPT ist es möglich die Messergebnisse vollständig zu reproduzieren. Das dem CPT zugrunde liegende Modell wurde mit analytischen Betrachtungen untermauert, sowie mit Computersimulationen einer eindimensionalen Fluidexpansion mit Ladungsseparation und zweidimensionalen, relativistischen Particle-In-Cell (PIC) Simulationen verglichen.

An o.g. Lasersystemen wurden des weiteren Experimente mit einer Verformung des Laserstrahlprofils auf der Vorderseite und dessen Auswirkung auf die Ionenbeschleunigung von der Folienrückseite durchgeführt. So konnte gezeigt werden, dass ein elliptisch geformter Laserfokus in einem elliptisch geformten Protonenstrahl resultiert. Diese Laserstrahlprofileinprägung nimmt mit zunehmender Dicke der Targetfolie ab. Die gleichzeitige Messung der Quellgröße der Protonenstrahlen mit Hilfe der mikrostrukturierten Folien führte zur Erkenntnis, dass der Elektronentransport bei 50 Mikrometer dicken Folien im wesentlichen durch Kleinwinkelstreuung bestimmt wird, die bei 13 Mikrometer dünnen Folien jedoch vernachlässigbar ist. Zur Interpretation der Messungen wurde der Sheath-Accelerated Beam Raytracing for IoN Analysis code (SABRINA) Code entwickelt, der aus einem gegebenen Laserstrahlprofil unter Berücksichtigung der Kleinwinkelstreuung das Intensitätsprofil des Protonenstrahls berechnet und die experimentellen Ergebnisse reproduziert. Die beobachtete, unerwartet große Emissionszone bei dünnen Folien ist höchstwahrscheinlich auf Rezirkulation der Elektronen in der Targetfolie zurückzuführen.

Zur weiteren Optimierung laserbeschleunigter Protonenstrahlen wurden Experimente mit verschiedenen Targetgeometrien durchgeführt. So ergab die Analyse der RIS-Daten von Experimenten mit neuartigen, konusförmigen Targets mit flacher Rückseite am TRIDENT bei moderaten Intensitäten von 10^{19} W/cm^2 mit 20 J in 600 fs eine nahezu zweifache Erhöhung der Maximalenergie der beschleunigten Protonenstrahlen, eine vierfach bessere Konversionseffizienz von Laserenergie in Ionenstrahlenergie sowie eine 13-fach höhere Ionenzahl über 10 MeV im Vergleich zu Daten von flachen Folien.

Interpretationen der Messungen der energieabhängigen Quellgröße und Divergenz und PIC-Simulationen zeigen eine glockenförmige Elektronenschicht auf der Folienrückseite als Verursacher der Divergenz laserbeschleunigter Ionenstrahlen. Diese kann durch geometrische Verformung der Targetfolie kompensiert werden. Erste Ergebnisse zur Kollimation von laserbeschleunigten Protonenstrahlen wurden am Z-Petawatt Laser erzielt.

Die Einkopplung von lasererzeugten Ionenstrahlen in konventionelle Beschleunigerstrukturen erfordert eine Separation der mit den Ionen propagierenden Elektronen. Dies kann durch einen Dipolmagneten erfolgen, wie in Experimenten am Z-Petawatt gezeigt werden konnte. In derselben Experimentkampagne konnte erstmals der kontrollierte Transport sowie die Fokussierung von laserbeschleunigten MeV-Protonen demonstriert werden. Hierzu wurden Miniatur-Quadrupollinsen, basierend auf Permanentmagneten mit Feldgradienten bis zu 500 T/m eingesetzt. Mit diesem Aufbau konnten 10^6 Protonen mit einer Energie von 14 MeV reproduzierbar auf eine Strahlgröße von ca. 300×200 Quadratmikrometer im Abstand von 50 cm von der Quelle fokussiert werden. Diese Kollimationsmethode und mögliche Energieselektion entkoppelt die relativistische Laser-Protonenbeschleunigung von der Strahlführung und der Fokussierung und erlaubt so erstmals beide Sektionen separat zu optimieren. Die Verwendung von Ionenlinsen ist ideal geeignet zur Anwendung an der nächsten Generation von hochrepetierenden Hochenergie-Kurzpulslasersystemen.

Die Ergebnisse der Arbeit führten zu einem verbesserten Verständnis der Protonenbeschleunigung durch Hochintensitätslaserbestrahlung dünner Metallfolien und ihrer Anwendungen [1–13]. Die Ergebnisse, speziell zum Transport und zur Fokussierung, haben einen weiteren Schritt zur breiten Anwendung laserbeschleunigter Ionen in verschiedensten Gebieten wie der Beschleunigerphysik, Trägheitsfusion, Astrophysik oder Strahlentherapie beigetragen.

Abstract

The irradiation of micrometer-thin metal foils by modern high-energy short-pulse lasers with intensities above $10^{18}\,\text{W}/\text{cm}^2$ leads to, amongst other things, the acceleration of ion beams with energies in the range of mega-electron-volts (MeV). Initially, the laser pulse accelerates electrons to relativistic energies, which then propagate through the foil. As soon as the electrons leave the foil's rear side, an electric field with a field strength of about $10^{12}\,\text{V}/\text{m}$ is generated. This effectively accelerates adsorbed protons from the foil surface in direction of the target normal. The quasi-neutral beams generated in such a manner consist of more than 10^{12} protons in a short, picosecond duration pulse. Possible applications are the diagnostics of dense plasmas, the utilization as compact injectors for particle accelerators, the energy generation by fast ignition in inertial fusion energy, as well as a potential utilization in cancer radio-therapy with ion beams.

Laser-accelerated ion beams exhibit some beam properties that are superior to ion beam properties from conventional ion sources. This motivates their application as a next generation ion source. Until today, though, there is no complete model for the laser ion-acceleration that can be used for estimates of all beam parameters. However, the development of applications requires an accurate knowledge of the space and momentum distribution (phase space) of the ions, as well as the best-possible modeling of the acceleration process. Therefore the measurement technique of "Radiochromic film Imaging Spectroscopy (RIS)" has been developed. The dosimetry films needed for RIS have been absolutely calibrated for protons at the tandem accelerator at the Max-Planck-Institut für Kernphysik in Heidelberg, Germany.

Furthermore, RIS uses the method of ion beam manipulation by micrometer-sized deformation of the foil surface. A technique has been developed to insert equidistant, micrometer-sized grooves (distance either 3, 5 or 10 micrometer) on the surface of thin foils with thicknesses from 5 to 50 micrometers. This is done by ultra-high precision-chipping of a carrier material, followed by electro-plated deposition of the foil and subsequent etching of the carrier material.

The micro-structured foils have been successfully used in experiments at the Petawatt High Energy Laser for Heavy Ion eXperiments (PHELIX) at the GSI Helmholtzzentrum für Schwerionenforschung GmbH (Darmstadt, Germany in March 2006), in two experimental campaigns at the TRIDENT laser at Los Alamos National Laboratory (New Mexico, USA, May 2005 and April 2006), at the 100 TW laser at the Laboratoire pour l'Utilisation des Lasers Intenses (École Polytechnique, Palaiseau, France, June 2006) and at the Z-Petawatt laser at Sandia National Laboratories (New Mexico, USA, December 2007). The data analysis not only confirms the findings obtained in earlier experiments, but additionally leads to conclusions about the electric fields driving the acceleration.

The results obtained with RIS have been considered in the development of the Charged Particle Transfer (CPT) code, that can be used for a three-dimensional simulation of the ion-acceleration from the rear side of the foil. CPT can fully reproduce the measured data. The underlying model in CPT has been confirmed by analytical examinations, computer simulations of a one-dimensional fluid expansion with charge separation and two-dimensional, relativistic Particle-In-Cell (PIC) simulations.

In addition, experiments on the action of a shaped laser beam profile at the target front side on the ion acceleration from the foil's rear side have been performed at the above-mentioned laser systems. It could be shown, that an elliptically shaped laser focus results in an elliptically shaped proton beam. Moreover, the laser beam profile impression becomes weaker with increasing target foil thickness. The simultaneous measurement of the proton beam source size by the use of the micro-structured foils lead to the conclusion that the electron transport in 50 micrometer thick foils is basically determined by small-angle scattering but is negligible for 13 micrometer thin foils. For the interpretation and reproduction of the experimental results the Sheath-Accelerated Beam Ray-tracing for IoN Analysis code (SABRINA) has been developed. This code calculates the intensity profile of the proton beam for a given laser beam profile, under consideration of small-angle scattering. The observed, unexpectedly large emission zone at thin foils is most likely the result of re-circulating electrons in the target foil.

Experiments with different target geometries have been performed for a further optimization of laser-accelerated proton beams. The RIS-data analysis from experiments with novel, cone-shaped targets with a flat rear side at TRIDENT with moderate intensities of 10^{19} W/cm^2 with 20 J in 600 fs showed a nearly two-fold increase of the maximum energy of the accelerated proton beams, a four-fold better conversion efficiency of laser energy to ion beam energy as well as a 13-fold higher ion number above 10 MeV compared to data from flat foils.

The interpretation of measurements of the energy-dependent source size and divergence and PIC simulations evidence a bell-shaped electron sheath at the foil's rear side as the originator of the divergence of laser-accelerated ion beams. The divergence could be compensated by geometrical deformation of the target foil. First experiments on the collimation of laser-accelerated proton beams have been obtained at Z-Petawatt.

The injection of laser-accelerated ion beams in conventional accelerator structures requires a separation of the co-propagating electrons and protons. This can happen by a dipole magnet, as could be shown in experiments at Z-Petawatt. During the same experimental campaign the first controlled transport and focusing of laser-accelerated MeV-protons could be demonstrated. For that purpose miniature quadrupole-lenses, based on permanent magnets with field gradients up to $500\,\text{T/m}$ have been utilized. 10^6 protons with an energy of $14\,\text{MeV}$ could be reproducibly focused to a beam spot of about 300×200 square micrometers with this set-up, in a distance of 50 cm from the source (see image on the cover). This collimation method and potential energy-selection decouples the relativistic laser-proton acceleration from the beam transport and focusing, paving the way to optimize both separately. The use of ion lenses is perfectly applicable for upcoming high-energy, high-repetition rate, short-pulse laser systems.

The results of this work have lead to a better understanding of proton-acceleration by high-intensity laser-irradiation of thin metal foils and their application [1–13]. The results, in particular the transport and focusing, have taken a further step towards a broad application of laser-accelerated ions in a large variety of fields like accelerator physics, inertial fusion energy, astrophysics and radiotherapy.

Contents

1 Introduction   **1**
    1.1 Overview: Laser-accelerated MeV ion beams 4
    1.2 Thesis structure 10
    1.3 Experimental campaigns 10

2 Relativistic laser-matter interaction and ion acceleration   **13**
    2.1 Single electron interaction 14
       2.1.1 The ponderomotive force 16
    2.2 Plasma interaction at the target front side 19
       2.2.1 Forward electron acceleration 20
    2.3 Laser-ion acceleration 25
       2.3.1 Fast-electron transport in dense matter 27
       2.3.2 Target Normal Sheath Acceleration - TNSA 34
    2.4 Expansion models 37
       2.4.1 Plasma expansion model 38
       2.4.2 Two-dimensional Particle-In-Cell (PIC) simulation 50

3 Experimental method   **61**
    3.1 General set-up and laser systems 61
    3.2 Ion beam detectors 63
    3.3 RadioChromic Film – RCF 64
       3.3.1 Film composition 65
       3.3.2 Radio-chemical reaction 66
    3.4 RCF calibration for protons 67
       3.4.1 Scanner calibration 68
       3.4.2 RCF sensitive layer calibration 70
       3.4.3 Energy deposition and dose rate sensitivities 78
    3.5 RCF Imaging Spectroscopy 81

		3.5.1	Opening angle	82
		3.5.2	Micro-grooved target foils	83
		3.5.3	Spectral reconstruction of laser-accelerated protons	85

4 Proton-acceleration experiments — 89

4.1 Typical parameters of TNSA-protons 89
4.1.1 Energy spectra of laser-accelerated protons 90
4.1.2 Comparison with expansion models 93
4.1.3 Energy-resolved opening angle 96
4.1.4 Energy-resolved source sizes 97
4.1.5 Beam emittance 99
4.2 Laser beam-profile impression on laser-accelerated protons 102
4.3 Beam optimization by target geometry 105
4.3.1 Cone-shaped target front side 105
4.3.2 Beam smoothing due to target thickness 110
4.3.3 Large scale curvature 112
4.4 Beam control with magnetic fields 118
4.4.1 Magnetic electron removal 118
4.4.2 Transport and focusing with quadrupole magnets 122

5 3D proton expansion model — 127

5.1 Sheath-Accelerated Beam Ray-tracing for IoN Analysis code - SABRINA 128
5.1.1 Application: Electron transport in solids 130
5.1.2 Discussion 134
5.1.3 Conclusion 135
5.2 3D-model based on flow characteristics 137
5.2.1 Essential physics of flow expansion 137
5.2.2 The Charged Particle Transfer code - CPT 144
5.2.3 Transfer function derived from a fluid approach 147
5.2.4 Transfer functions for TNSA-protons 148
5.2.5 Numerical implementation 150
5.3 Reconstruction of experimental data 151
5.4 Expansion dynamics 154
5.5 Summary 157

6	**Outlook**	**159**
	6.1 Further optimization and control	159
	6.2 Generation of high-energy density matter	165
A	**Plasma Simulation Code - PSC**	**169**
	A.1 Governing equations	169
	A.2 Numerical implementation	170
	A.3 Hard- and software installation	172

Publications	**201**
Acknowledgements	**205**

Chapter 1

Introduction

Ever since lasers were invented in 1960, their peak power and peak intensities have steadily increased. In recent years, the invention of the Chirped Pulse Amplification (CPA) technique by Strickland and Mourou [14] allowed for the construction of laser systems, able to create Petawatt (10^{15} W) laser pulses with energies on the order of 1000 J, wavelengths $\lambda_L = 1\,\mu$m and pulse durations below one picosecond. When focused to micron spot-sizes with adaptive optics, electromagnetic intensities up to 10^{21} W/cm^2 can be reached, opening up a new research field called *high-field physics*.

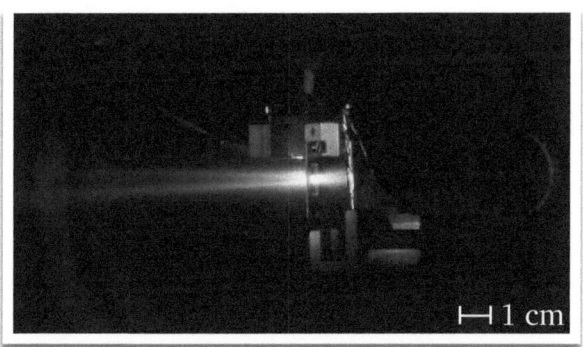

Fig. 1.1: Photograph of a high-intensity, short-pulse laser-matter interaction experiment. The (invisible) laser pulse with intensity $I_L > 10^{18}$ W/cm^2, wavelength $\lambda_L = 1\,\mu$m, pulse duration $\tau_L < 1$ ps irradiates a $10\,\mu$m thin gold foil target, mounted in an aluminum frame, from the left side of the image. An ion detector, wrapped in aluminum foil for shielding, has been placed behind the foil. The image was taken by K.A. Flippo during an experimental campaign at the TRIDENT laser facility at Los Alamos National Laboratory, NM, USA.

By irradiating solid matter with intense laser pulses, the matter almost instantly transforms to a high-density plasma state. Figure 1.1 shows a photograph of the laser-matter interaction, taken with a conventional, digital Single-Lens Reflex (SLR) camera by K.A. Flippo during an experimental campaign at the TRIDENT laser facility at Los Alamos National Laboratory, New Mexico, USA. The (invisible) laser pulse irradiates a $10\,\mu$m thin gold foil target, mounted in an aluminum frame, from the left side of the image. The laser creates a hot, dense plasma on the front side of the foil. The plasma emits radiation in a broad range of the electromagnetic spectrum, a part of the radiation is in the visible range and could be observed by the SLR in form of the intense, white light. Non-linear interaction creates higher harmonic radiation, e.g. the laser is frequency-doubled to $\lambda = 527$ nm, visible as the green light in the image. For later times, long after the femtosecond laser pulse has ended, a large part of the foil has been transformed to a plasma, that has expanded into vacuum thereby emitting the white-colored light recorded by the SLR.

During the laser irradiation, the force exerted by laser pulses with intensities above 10^{18} W/cm^2 accelerates electrons to energies of million electron-volts (MeV) in distances of a few microns. The electron energy becomes greater than the rest mass, hence the laser-electron interaction becomes relativistic. The acceleration gradient is a thousand times greater than in conventional, radio-frequency-based accelerators. This regime of relativistic laser-electron acceleration also leads to the acceleration of ions, with tremendously different characteristics compared to ions emitted from nanosecond laser-plasmas. The ions form a highly laminar, collimated beam with energies up to ten's of MeV. These unique features may allow laser-ion sources to be useful someday for cancer radiotherapy or as accelerators for nuclear physics research. They might also be applied to ignite controlled thermonuclear fusion for energy production.

Due to these prospects, there is lots of scientific activity in the field. However, the relativistic laser matter interaction and ion acceleration is very complex and the understanding of the physics is still at a premature stage. This is reflected in the literature, nearly each week another publication appears with "new" findings about laser matter interaction, electron or ion acceleration. Furthermore, the non-linear, relativistic, collective interaction strongly limits analytical approaches. Instead, computer simulations are used to get an insight into the laser matter interaction and ion acceleration. However, the use of modern computer codes is as complex as an experiment, it requires large computing clusters and generates huge amounts of data from where the relevant physical processes have to be extracted. Even today, the optimum conditions for reliable, efficient and energetic laser ion acceleration still have to be worked out. Nevertheless, the basic mechanisms driving the ion acceleration have been found. The basic model is as follows: First a plasma is created on the target front side by the unavoidable pre-pulse of the laser. The strong electromagnetic field in the focal spot of the short main pulse accelerates electrons by various mechanisms to MeV energies. The electrons are able to follow the quick oscillations of the laser field (period $T \approx 3\,\text{fs}$), whereas the ions remain stationary. The displacement leads to space charge fields on the same order of magnitude as the laser field. Additionally, copious amounts of electrons are accelerated towards the solid target and penetrate it. As soon as they enter the vacuum at the rear side, a strong electric field is created. The field ionizes the atoms at the surface, which are then accelerated in target normal direction.

As simple as this model is, as complex are the details, e.g. how does the acceleration scale with the various laser and target parameters? First scaling laws [15–17] have been found, that are able to explain some maximum ion energies obtained at different laser systems worldwide. What has to be done, though, is to further understand and better control the acceleration of ions by intense lasers, since without that the proposed applications cannot be realized.

The next section tries to give an overview of the field of relativistic laser ion acceleration, in order to show what has been investigated and how it affects the acceleration of ions from solid matter, such as this work could shed some light in the expansion dynamics of laser-accelerated ions, the focusability of protons by magnetic ion lenses and the role of the laser beam profile impression and target thickness impact on laser-accelerated protons. After the overview, the structure of the thesis is explained in section 1.2. It is followed by an overview about the experimental campaigns where the author has participated in section 1.3.

1.1 Overview: Laser-accelerated MeV ion beams

Since its first discovery in the year 1999 [18–24] the field of laser-ion-acceleration from thin foil targets got strong scientific attention. The outstanding features of laser-accelerated ion beams promise their use in a large variety of applications. The beams can be used as a diagnostic tool in basic plasma research, e.g. for the diagnostics of electromagnetic fields in dense plasmas with picosecond time resolution [25–28] or for the creation of high-energy density (HED) matter [29,30]. Looking further into the future, laser-accelerated ions could be applied as compact particle accelerators [31–34], as a driver for neutron production [35,36], for radioisotope generation [37–40], for table-top nuclear physics [41], for the generation of intense K_α x-rays [42], for Inertial Fusion Energy in the case of Proton Fast Ignition [43,44] or even for medical applications as a compact radiotherapy system for tumor treatment [45–48].

Without special target cleaning techniques [5, 6, 8, 49, 50] the predominantly accelerated ion species are protons from contamination layers on the target surface [51]. Although it was not clear from the beginning whether the protons originate from the front or from the rear side, it was observed that the physical mechanism driving the acceleration was very robust and reproducible, always leading to a relatively collimated beam pointing in the target normal direction. The reason for the obfuscation is, that on the front side the ions can be accelerated to high energies in laser direction due to the laser-generated charge-separation at the critical density of the plasma. At the rear side a dense sheath of energetic electrons, that have propagated through the target, creates an electric field on the order of TV/m, which ionizes and accelerates the atoms at the rear side. This acceleration scheme is known as the *Target Normal Sheath Acceleration (TNSA)* [52].

Very soon there was clear evidence for the rear-side emission to be the source of MeV-energy protons [53–56] and heavy ions [49,57]. Even two years before the first discovery Tatarakis *et al.* [58] observed a hot plasma expansion from the rear side of a thick ($d > 140\,\mu$m) plastic foil, irradiated by an $I = 10^{19}\,$W/cm^2 laser pulse. The creation of a charge-separation sheath with an electric field strength above $5 \times 10^{11}\,$V/m at the rear side is already mentioned there. However, they did not measure accelerated protons hence the honor of the discovery of MeV-ion acceleration belongs to the aforementioned authors.

Kaluza *et al.* [59] have shown, that the level of the laser-pre-pulse and target thickness determine if the most energetic protons originate from the front or the rear side.

An experimental comparison of front versus rear side acceleration was done by Fuchs et al. [60, 61]. Their findings are supported by computer simulations by Sentoku et al. [62]. The conclusion is, that for the laser intensities ($I = [10^{18}, 10^{20}]\,\text{W/cm}^2$) as well as the target thicknesses ($d = [5, 50]\,\mu\text{m}$) discussed in this thesis, the rear-side acceleration produces higher energetic and better collimated ion beams more efficiently.

The ions are accelerated by a strong electric field created by a dense sheath of hot electrons in the vacuum region. The transverse dynamics of the hot electron sheath at the target's rear side has been investigated in refs. [26, 36, 57, 63]. With the help of computer simulations McKenna et al. [63] found, that the electron sheath at the rear side spreads along the surface with a velocity $v \approx 0.75\,c$, where c is the speed of light in vacuum. The radial expansion leads to strong electric fields on the order of GV/m even millimeters away from the laser focus region. In most experiments the target foils have widths and heights on the order of a few millimeters. At the foil edges an enhanced electric field is created, leading to ion emission as well. These edge-emittted ions have much less energy than the ions created in the central sheath region. The edge acts as a cylindrical lens and focuses these ions to a line-shaped beam, similar to the ion acceleration from wire-targets [54, 64].

The electrons forming the sheath at the rear side can be accelerated back into the target by the electric field. They start re-fluxing back and forth both sides of the foil, that enhances the sheath density and thereby enhances the electric field driving the acceleration [65]. The field strongly peaks at the ion front, due to the charge separation by the electrons propagating in front of the ions. The field has been measured by time-resolved proton radiography [26].

Ion beams in conventional accelerators are all of the same kinetic energy. On the contrary, laser-accelerated ions usually exhibit a whole spectrum, i.e., the particle number decays exponentially with increasing energy. The spectrum is not infinite but exhibits a cut-off with a sudden drop to zero at the maximum energy. The particle number can be above 10^{13} particles in total [66]. The longitudinal acceleration dynamics and the resulting spectra have been discussed by Mora et al. in refs. [67–71]. Experimentally determined proton spectra will be discussed in section 4.1.2.

Both, the number dependence on energy and the opening angle of the ions, depend on the energy. By measuring the source size of the protons [3, 57, 72–77] it was found that high-energy protons originate from a much smaller source than lower energetic ones. Hence, the source size of the protons is energy-dependent, too. In fact, the decreasing opening angle with increasing energy is not due to a reduction in beam divergence with energy, but due to the decreasing source size [18, 56, 72]. Protons with higher energy even have little higher transverse momentum than the lower energetic ones. This has been already mentioned by Wilks *et al.* in the original publication explaining TNSA [52].

The divergence of the ion beam can be reduced by curving the whole target. After the first proof-of-principle experiments [54], the focusability of protons by spherically curved targets has been investigated refs. [29, 30, 78–80] in the context of high-energy density matter generation. A different approach for proton focusing was the invention of a laser-triggered microlens device [81] and its further development using a sophisticated target assembly [82]. In the framework of this thesis, a more reliable and straightforward approach by using novel permanent magnet mini-quadrupole devices has been applied to transport and focus laser-accelerated protons, allowing for a reproducible beam manipulation. The results have been published in ref. [1].

One striking feature of laser-accelerated ion beams is their high degree of laminarity. This is reflected in the observation, that sub-micrometer-sized corrugations of the target rear surface are imprinted in the beam and are imaged over centimeter distances [54]. This effect can be used to determine the beam quality in terms of emittance [72, 83]. The smaller the emittance is, the higher is the beam laminarity and focusability. Compact, modern accelerators, e.g. the Heidelberg Ion Beam Therapy Center (HIT) at Heidelberg, Germany, have a transverse emittance on the order of $1\,\mathrm{mm\,mrad}$ [84, 85]. In contrast to that, the transverse emittance of laser-accelerated protons is less than $10^{-2}\,\mathrm{mm\,mrad}$ [9, 54, 72, 83, 86] and below $10^{-1}\,\mathrm{mm\,mrad}$ for heavy ions [57].

The theoretical explanation for the imaging-effect of micro-corrugations has been given by Ruhl *et al.* [87, 88] with particle-in-cell (PIC) computer simulations and a numerical code based on transfer functions. In the framework of this thesis the model was further developed and compared to experimental data and simulations.

By coating the rear side with a thin ($d < 100\,\mathrm{nm}$) plastic film a very smooth rear surface can be obtained, resulting in a smooth proton beam [54, 76] without significantly decreasing the acceleration efficiency. A thicker rear side coating, though, disturbs the electron transport from the front to the rear side due to the different electrical conductivities of the substrate and the coating, and results in substantially lower conversion efficiency as well as corrugated ion beams [54, 76, 89].

The atomic composition at the rear side can lead to spectral modulations (peaks), when the expanding beam does not only consist of electrons and protons but heavier ions as well. The heavy ions modify the electric field in front of them, leading to an accumulation of protons [56, 90, 91]. The ratio of the proton density at the rear side to the density of other contamination ions, e.g. carbon, determines the strength of these modulations [92]. In addition to that, the presence of some co-moving ions can also lead to dips in the spectrum [93–96].

A very thin (nm-sized) and small layer at the target's rear side was proposed to lead to quasi-monoenergetic proton spectra [97, 98] and indeed found in experiments with protons [99] as well as with carbon ions [100, 101]. However, recent investigations have shown that the spectral peak found in ref. [99] is not due to the thin proton layer at the rear side, but due to the density composition of the protons and heavier ions [102, 103]. The density composition effect has been used as well in experiments with heavy-water micro-droplet targets, leading to quasi-monoenergetic deuteron beams [94].

The conversion efficiency (laser energy to proton beam energy) as well as the maximum proton energy both scale proportional to the laser intensity $I^{1/2}$ as found by Clark et al. [20] for protons and by Hegelich et al. [104] for heavy ions. A detailed study on the scaling of proton energy and conversion efficiency - both depending on laser intensity and target thickness - has been published by Fuchs et al. [15]. This study was extended to higher laser intensities by Robson et al. [16] at the Rutherford Appleton Laboratories VULCAN petawatt laser, where conversion efficiencies about 10 % were obtained. This high conversion efficiency has been already obtained in the very first experiments at the NOVA Petawatt laser [21, 22, 66, 105]. The results imply that not the most intense, but the most energetic lasers are able to accelerate ions to highest energies. The highest maximum proton energies of approximately 60 MeV reached to date have been found at the two high-energy petawatt lasers NOVA Petawatt [21] and VULCAN Petawatt [16]. Robson et al. [16] further showed that the maximum energy still scales as $I^{1/2}$ up to at least $I_{\max} = 6 \times 10^{20}$ W/cm^2. However, it was necessary to modify the model proposed in Fuchs et al. [15] to explain the weaker-than-expected maximum energies with higher intensities. According to ref. [15], a significantly higher proton energy, e.g. 100 MeV, should be obtained by simply taking longer pulses in the picosecond range, with more energy to keep the intensity constant. On the contrary, ref. [16] reports on a relatively constant maximum proton energy with increasing pulse duration from 1 ps to 10 ps while keeping the laser intensity constant. To date there is no clear explanation for the discrepancy. Reasons could be possible pre-heating of the rear side or a stronger-than-expected spread of the sheath in transverse direction during the acceleration, leading to cooling of the electrons. The comparison of the experiments conducted during this thesis and the scaling law from ref. [15] will be done in section 4.1.2.

Besides the laser intensity and target thickness, there are many more parameters that influence the acceleration. A scaling of the maximum proton energy with the pre-plasma scale length was first done by Kaluza *et al.* [59], showing that optimum ion acceleration depends on the laser pre-pulse duration and target thickness. An optimum pre-plasma also leads to a smooth ion beam profile [106].

The pre-pulse generates a shockwave traveling into the solid matter. The shockwave velocity depends on the pre-pulse intensity and it is a few times faster than the sound velocity c_s, which is e.g. $3\,\mu\mathrm{m/ns}$ for gold at normal conditions [107]. Hence the pre-pulse duration of the laser should be less than a few nanoseconds to maintain an undisturbed target rear side. Otherwise, the shockwave breaks out at the rear side and creates a plasma at the rear side. The negative influence of a large-scale length ($l_s = 100\,\mu\mathrm{m}$) plasma on the proton maximum energy was demonstrated by Mackinnon *et al.* [53]. The protons are emitted as a ring-like proton beam with low energy [3, 108]. Contrary to that, small-scale plasma gradients ($l_s < 5\,\mu\mathrm{m}$) seem to have very little influence on the maximum energy [109].

The extreme case of zero pre-pulse would result in zero pre-plasma, that is less efficient for hot electron generation. Thus, there is an optimum pre-pulse level leading to efficient TNSA [34, 110–112].

Under certain pulse contrast conditions, when the pre-pulse generated shockwave has just reached the target's rear side, the surface will be deformed by the breakout, resulting in a change of the ion beam pointing out of the target normal direction [107, 113–115].

Furthermore, a significant reduction of the pre-pulse allows the use of ultra-thin target foils and therefore very efficient proton acceleration [116] up to 10 MeV for 50 nm thin foils, compared to a maximum energy of 1 MeV without further pre-pulse suppression and a $5\,\mu\mathrm{m}$ foil [117]. Going to even further pre-pulse elimination and very short laser pulses ($\tau_L = 65\,\mathrm{fs}$), the laser pulse can efficiently heat very thin foils as a whole, and the TNSA-effect can accelerate protons in forward as well as in backward direction with equal efficiency and energy [118].

After the discussion of the pre-pulse influence on proton-acceleration, and the scaling with main pulse energy and intensity, it should be noted how further parameters of the main pulse influence the proton acceleration: The pulse can be either spatially varied (i.e., a modulation of the transverse intensity distribution from a tightly focused Gaussian spot to a broader distribution), it can be temporally varied (e.g. by employing two main pulses separated by a certain time delay) or the polarization of the main pulse can be changed from the usual linear polarization to elliptical or circular polarization.

The influence of the main pulse beam profile on proton-acceleration was first discussed by Fuchs *et al.* [76], showing that a deformed (i.e., not radially symmetric) laser pulse profile imprints in the proton beam profile. It was found that an elongated laser intensity profile creates an electron sheath that follows the laser beam topology. The sheath then accelerates protons with a transverse beam profile that resembles features of the laser beam profile [11, 119]. This laser beam imprinting was further investigated in experiments during this thesis. In contrast to the experiments by Fuchs *et al.*, the source size of the protons was simultaneously measured with the help of micro-grooved targets. Additionally a target thickness scan as well as a laser beam spot size scan was performed. The results will be shown in chapter 4 and have been published in refs. [2, 3].

The other two options, the change of the laser polarization or a double pulse configuration, are to date only considered theoretically by computer simulations and analytical estimates. It is proposed that two temporally separated laser pulses should lead to spectral peaks as well as an artificially collimated electron transport through the target [120, 121]. The option of using circularly polarized laser beams is discussed in refs. [122–125]. It might generate high-energy, low-divergence ion beams with a non-exponential spectrum. However, this scheme requires high-intensity pulses with very high pulse contrast, making it very difficult to realize experimentally.

In conclusion, this overview has summarized the state-of-the-art in short-pulse laser-ion acceleration. A lot of experiments have been done, but there are still a lot of open questions with respect to the optimization and full control of laser-ion acceleration. Some of them will be considered in the next chapters.

1.2 Thesis structure

The thesis is divided in four major parts. The first part explains the theoretical models explaining the relativistic laser-matter interaction and ion acceleration in chapter 2. It starts with the interaction of a plane electromagnetic wave with a single electron. Then the generation of hot electrons at the target front side, the electron transport through the target and the subsequent acceleration of ions off the rear surface are discussed. The chapter ends with the presentation of expansion models that were used to explain the experimental results. As outlined below, the nature of the experiments limits the number of data points. Hence a detection method had to be developed that could get as much information about the accelerated beam as possible, within a single measurement. This lead to the development of radiochromic film imaging spectroscopy. The radiochromic film detector and the measurement technique are explained in chapter 3. The main part of the thesis is the experimental investigation on optimization possibilities and further control of laser-accelerated protons, presented in chapter 4. With respect to possible applications, the most important aspect that needs better control is the beam divergence. Various options like shaping the proton beam by shaping the laser focal spot, changing the target geometry or the application of external magnetic fields have been tested. The explanation of the results required the development of proton beam expansion models. These are outlined in chapter 5. One model has been used able to explain the laser beam imprinting and target thickness impact on laser-accelerated protons, whereas another model could be applied to fully reconstruct the measured beam. The close affinity to plasma expansion models will be discussed there, too. Once the divergence is under control, a huge variety of applications can be realized. A specific application, that is of great interest in basic plasma research, inertial fusion energy and astrophysics, is the generation and diagnostics of high-energy density matter. Laser-accelerated protons can be used for the preparation of such an extreme state of matter, as will be shown in chapter 6. Besides that, an outlook on further optimization is given.

1.3 Experimental campaigns

The experiments have been carried out at fairly large laser systems in Europe and the USA. Each user group can access the facility for a limited amount of time, usually a few weeks, which is called beamtime. Since a beamtime is scheduled to a group of users, each campaign usually covers more than one research topic. The repetition rate (between 20 min. and two hours) of the Nd:glass laser amplifiers only allows for a few laser discharges ("shots") a day. Hence the number of measurements is limited to a few shots per research topic.

The author has participated in experimental campaigns at five lasers systems. Two experimental campaigns have taken place at the TRIDENT laser facility [126] at Los Alamos National Laboratories (LANL), Los Alamos, New Mexico, USA in the years 2005 and 2006. The group of experimenters consisted of J.A. Cobble, K.A. Flippo, D.C. Gautier, S. Letzring, J.C. Fernández and B.M. Hegelich from LANL, J. Schreiber from the Ludwig Maximilians Universität (LMU) München, Germany and the author. The topics of one campaign were the repetition of the acceleration of quasi-monoenergetic heavy ions, heavy-ion acceleration from targets cleaned by intense laser-ablation, the investigation of the acceleration of heavy ions buried deep below the surface and the laser beam imprinting and target thickness effect on protons. The other campaign was about the enhanced acceleration-efficiency by the use of flat-top cone targets and further investigations of the ion beam parameters by micro-structured targets. The results have been published in refs. [2–6, 17, 50, 100, 127].

Another campaign has been carried out at the 100 TW laser system [128] at the Laboratoire pour l'Utilisation des Lasers Intenses (LULI) at the École Polytechnique, Palaiseau, France in 2006. The goal of the experiment was the spatial resolved measurement of the electric field on the rear surface of a laser irradiated thin foil by field ionization, a further investigation of the laser beam imprinting effect and the application and calibration of a newly developed Thomson parabola ion detector. The group consisted of E. Brambrink and P. Audebert from LULI, J. Schreiber from LMU, K.A. Flippo, D.C. Gautier and B.M. Hegelich from LANL, M. Geißel from Sandia National Laboratories (SNL), Albuquerque, New Mexico, USA, K. Harres, F. Nürnberg and the author from the Technische Universität Darmstadt (TUD), Germany. The results are published in refs. [2, 8].

The very first laser-proton-acceleration experiments at the GSI Helmholtzzentrum für Schwerionenforschung GmbH at Darmstadt, Germany have been carried out during the commissioning of the Petawatt High-Energy Laser for heavy Ion eXperiments (PHELIX) [129] in the year 2006 as well. The group consisted of E. Brambrink from LULI, J. Schreiber from LMU, B. Zielbauer and K. Witte from GSI, K. Harres, F. Nürnberg, M. Roth and the author from TUD. With the front-end and the preamplifier of the PHELIX the acceleration of MeV-protons from thin gold foils could be demonstrated. Some results are published in ref. [3].

In 2007, the author has participated in an experimental campaign about the diagnostics of high-energy density (HED) matter by spectrally resolved x-ray Thomson scattering (XRTS) at the Janus laser at Jupiter laser facility, Lawrence Livermore National Laboratory (LLNL), Livermore, California, USA. The group consisted of S.H. Glenzer, A. Kritcher, H.J. Lee, P. Neumayer and D. Price from LLNL, G. Gregori from the Rutherford Appleton Laboratory, Didcot, UK, E. García Saiz from Queen's University of Belfast, UK, A. Pelka and M. Roth from TUD. Although the topic was not about ion-acceleration, XRTS is a very promising candidate for the diagnostics of HED matter generated by the irradiation of a second foil by laser-accelerated protons. Details about this application can be found in chapter 6. Some results of the experimental campaign will be published in ref. [7].

The last experimental campaign has been carried out at the Z-Petawatt [130] laser at SNL in december 2007. The experiments about the control of laser-accelerated protons by externally applied magnetic fields and the focusing of protons by curved target foils have been carried out by M. Geißel, M. Kimmel, P. Rambo and J. Schwarz from SNL, K. Flippo from LANL, J. Schütrumpf from TUD and the author.

Chapter 2

Relativistic laser-matter interaction and ion acceleration

The laser pulses discussed in this thesis are able to create a state of matter just beginning to be explored. A laser pulse with an intensity I_0 above $10^{18}\,\text{W}/\text{cm}^2$ has a corresponding electric field amplitude of

$$E_0 = \sqrt{2I_0/\varepsilon_0 c} \approx 2.7 \times 10^{12}\,\text{V}/\text{m}, \qquad (2.1)$$

where ε_0 denotes the electric constant and c the speed of light in vacuum [131]. This field is almost an order of magnitude larger than the electric field of an hydrogen atom, which can be estimated by

$$E = \frac{e}{4\pi\varepsilon_0 a_B^2} \approx 5.1 \times 10^{11}\,\text{V}/\text{m}. \qquad (2.2)$$

The quantity $a_B = 4\pi\varepsilon_0 \hbar^2/m_e e^2 = 5.3 \times 10^{-11}\,\text{m}$ is the Bohr radius, with the Dirac constant $\hbar$, electron mass m_e and electron charge e. Hence the electron is no longer bound to the nucleus but it will oscillate in the laser's electromagnetic wave with relativistic velocity ($v \approx c$), which results in a relativistic mass increase exceeding the electron rest mass. At this point, the magnetic field of the laser pulse's electromagnetic wave comes into play, that changes the interaction physics and non-linear effects become important. The propagation of light also becomes dependent on the light intensity, resulting in non-linear effects as well.

In order to understand these phenomena, the motion of single electrons in an intense electromagnetic wave is considered in the next section. Following this rather academic introduction, section 2.2 deals with the interaction of an intense laser pulse with plasma since already the onset of a high-intensity laser pulse, with orders of magnitude less intensity, is able to create a plasma at the surface of a solid. The main part of the laser pulse then interacts with the plasma. Different absorption mechanisms in comparison to those in nanosecond laser-plasma interaction [132] are introduced. The absorption leads to strong electron acceleration into the solid, explained in sec. 2.2.1. The electrons in turn are able to accelerate ions from the target rear side. Details of this laser-ion acceleration are given in section 2.3. The introductory theoretical part then ends in section 2.4 with the presentation of plasma models used for the description of the expansion of laser-accelerated ions and their numerical realization.

For the main part of the next sections the references [127, 133–135] and references therein have been of great help and were used as a guide in the rich physics involved in intense laser-plasma interaction.

2.1 Single electron interaction

The laser pulse is an electromagnetic, linearly polarized wave propagating in z-direction, given as

$$\boldsymbol{E}(x,y,z,t) = E_0(t)\,e^{-i(\omega_L t - kz)}\,\boldsymbol{e}_x \tag{2.3}$$

$$\boldsymbol{B}(x,y,z,t) = B_0(t)\,e^{-i(\omega_L t - kz)}\,\boldsymbol{e}_y \quad \text{with } B_0 = E_0/c, \tag{2.4}$$

where $E_0(t)$ and $B_0(t)$ are the slowly varying[1] field amplitudes, ω_L the laser angular frequency, $k = \omega_L/c$ the laser wave vector and $\boldsymbol{e}_{x,y}$ are normalized vectors, both normal to the propagation direction $\boldsymbol{e}_z$, respectively. Both fields are connected via the third Maxwell equation $\nabla \times \boldsymbol{E} = -\partial \boldsymbol{B}/\partial t$ to $c\,\boldsymbol{B} = \boldsymbol{e}_z \times \boldsymbol{E}$.

The motion of a single electron in vacuum with charge e in presence of the light wave is described by the Lorentz force

$$\frac{\mathrm{d}\boldsymbol{p}}{\mathrm{d}t} = -e\,(\boldsymbol{E} + \boldsymbol{v} \times \boldsymbol{B}) = -e\left[\boldsymbol{E} + \boldsymbol{v} \times \left(\frac{1}{c}\boldsymbol{e}_z \times \boldsymbol{E}\right)\right], \tag{2.5}$$

where $\boldsymbol{p} = \gamma m_e \boldsymbol{v}$ and $\gamma = \left(1 + p^2/(m_e^2 c^2)\right)^{1/2}$ is the relativistic factor.

For non-relativistic velocities $v \ll c$ the force acting on the electron is given by the electric field only. The solution of the equation of motion leads to a harmonic

---

[1] The slowly varying envelope approximation assumes the temporal envelope $f(t)$ to be slowly varying relative to the laser cycle with frequency ω_L: $\mathrm{d}f/\mathrm{d}t \ll \omega_L f$.

oscillation in x-direction with the amplitude or electron quiver velocity

$$v_{\text{osc.}} = \frac{eE_0}{m_e\omega_L}. \tag{2.6}$$

For large electric field amplitudes $E_0 > 3.2 \times 10^{12}\,\text{V/m}$ the electron quiver velocity can approach c, that is for intensities $I > 1.37 \times 10^{18}\,\text{W/cm}^2$ according to eq. (2.1). Hence the laser-electron interaction is called relativistic if the dimensionless electric field amplitude

$$a_0 := \frac{eE_0}{m_e\omega_L c} = \sqrt{\frac{I_0\,[\text{W/cm}^2]\lambda_L^2\,[\mu\text{m}^2]}{1.37 \times 10^{18}\,\text{W/cm}^2}} > 1. \tag{2.7}$$

For the laser pulses discussed in this thesis a_0 is between one and 20; the oscillation velocity is close to the speed of light. Hence the magnetic component of the Lorentz force has to be taken into account when solving eq. (2.5). This can be done by changing to the vector potential $\boldsymbol{A}$ with the relations $\boldsymbol{E} = -\partial\boldsymbol{A}/\partial t$ and $\boldsymbol{B} = \nabla \times \boldsymbol{A}$, which is exercised in Ref. [133], p. 31. The scalar potential is zero in vacuum. Figure 2.1 shows the resulting trajectory (—) of an electron initially at rest in presence of the light wave. The normalized laser amplitude is $a_0 = 1$, the wavelength of $1\,\mu\text{m}$ leads to $k = 6.3 \times 10^6\,\text{m}^{-1}$. This corresponds

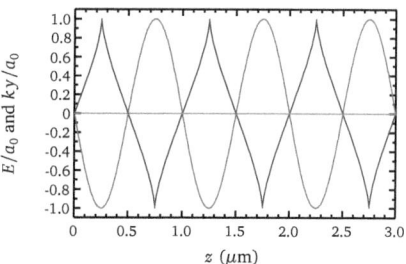

Fig. 2.1: Electron orbit (—) in a linearly polarized electromagnetic field with amplitude $a_0 = 1$. The electron oscillates in the plane of the electric field (—) with a wavelength of $1\,\mu\text{m}$ ($k = 6.3 \times 10^6\,\text{m}^{-1}$). The magnetic field (—) oscillates perpendicular to this plane and leads to a forward drift of the electron.

to an intensity $I = 1.37 \times 10^{18}\,\text{W/cm}^2$. The electric field (—) oscillates in the yz-plane, the magnetic field (—) oscillates in the xz-plane perpendicular to this plane. The magnetic field force $F = q\,v\,B \propto \boldsymbol{E}\cdot\boldsymbol{E}/c \propto a_0^2$ leads to a forward drift of the electron motion in z-direction, independent of the laser polarization. The corresponding velocity is

$$\boldsymbol{v}_D = \frac{a_0^2}{4 + a_0^2}\,c\,\boldsymbol{e}_z. \tag{2.8}$$

However, although the electron has changed its position, at the end of the laser pulse the electron velocity is zero again. Hence the electron does not gain energy from the laser, which is known as the Lawson-Woodward theorem [136].

2.1.1 The ponderomotive force

The solution presented above is only valid for a plane wave, i.e., an electromagnetic wave that is uniform in space and only slowly varying in time. Laser pulses in the real world are far from being that ideal: the tight focusing and short pulse duration create strong gradients in all directions over a few micrometers. The electromagnetic field distribution is not constant, but resembles, for example, a Gaussian shape. An electron accelerated in this field is pushed to lower intensity regions through the *ponderomotive force*. Although the ominous term ponderomotive force can already be found in literature of the 19th century [137], nowadays the ponderomotive force is defined as the low-frequency force fraction of a spatially inhomogeneous, high-frequency electromagnetic field on charged particles. It was derived in 1957 by Boot and Harvie [138], who showed that in a non-uniform radio-frequency field the oscillation center dynamics of a free electron is governed by a force that originates from second-order terms of the Lorentz-force (eq. (2.5)). For the non-relativistic limit ($v \ll c$), and in a one-dimensional case but with an intensity dependence in z-direction, the Lorentz-force becomes $\partial v_z / \partial t = -e/m_e\, E_z(z)$, where E_z is taken from eq. (2.3) but now with a z-dependence. This non-linear equation can be solved by Taylor-series expansion and by sorting by the orders of E_z. To lowest order the solution for a plane wave shown above is obtained. The second-order term gives a fast-oscillating term

$$\frac{\partial v_z^{(2)}}{\partial t} = -\frac{e^2}{m_e^2 \omega_L^2} E_0 \frac{\partial E_0(y)}{\partial y} \cos^2(\omega_L t - kz). \qquad (2.9)$$

Multiplying by m_e and taking the cycle-average yields the ponderomotive force:

$$\boldsymbol{f}_p = -\frac{e^2}{4 m_e \omega_L^2} \nabla (\boldsymbol{E} \cdot \boldsymbol{E}^*). \qquad (2.10)$$

Physically the electron is pushed away from the region of locally higher intensity, picking up a velocity $v \propto v_{\text{osc}}$, which is just the quiver velocity from above.

The relativistically correct equation of motion has been obtained by Bauer *et al.* [139] to

$$\boldsymbol{f}_p = -\frac{c^2}{\gamma} \left[\nabla m_{\text{eff}} + \frac{\gamma - 1}{v_0^2} (\boldsymbol{v}_0 \cdot \nabla m_{\text{eff}}) \boldsymbol{v}_0 \right], \qquad (2.11)$$

with the space- and time-dependent effective mass

$$m_{\text{eff}} = \left(1 + \frac{e^2 \boldsymbol{A} \cdot \boldsymbol{A}^*}{2m_e^2 c^2}\right)^{1/2} = \bar{\gamma}, \tag{2.12}$$

in a linearly polarized monochromatic wave in vacuum. The last term contains the cycle-averaged gamma factor $\bar{\gamma} = \sqrt{1 + a_0^2/2}$ [140]. For a circularly polarized wave $\bar{\gamma}$ equals $\sqrt{1 + a_0^2}$ [134, 139, 140], which is slightly different.

In the fully relativistic case the solution of the equation of motion is very complicated and has to be done numerically [139] or it has to be simplified for special cases [141], since the force is a nonlinear function of the electron's momentum and position. However, the energy the electron gains can be obtained by calculating the ponderomotive potential U_{pond} via $\boldsymbol{f}_p = -m_e \nabla U_{\text{pond}}$, which leads to $W_{\text{pond}} = \bar{\gamma} m_e c^2 + C$. The integration constant C is determined by the fact that in the non-relativistic limit the ponderomotive energy must devolve to $W_{\text{pond}} = m_e v_{\text{osc}}^2 / 4$, which is the case when $C = -m_e c^2$, as can be shown by insertion and Taylor-expansion of $\bar{\gamma}$. Therefore the resulting equation for the energy gained by the relativistic ponderomotive potential is:

$$W_{\text{pond}} = (\bar{\gamma} - 1) m_e c^2. \tag{2.13}$$

For an illustration of the complex interaction of an intense laser pulse interacting with free electrons, it has been simulated in two spatial dimensions and three momentum and field dimensions (2D3V) with the fully relativistic Particle-In-Cell (PIC) code PSC, which is explained in section A. The quadratic simulation box with $20\,\mu\text{m}$ length was divided into 1000×1000 cells. At $z = (1 \pm 0.01)\,\mu\text{m}$ 6000 electrons have been placed. The laser pulse with $\lambda = 1\,\mu\text{m}$ was modeled as a Gaussian in space and time with $3.3\,\mu\text{m}$ full width at half maximum (FWHM) in y-direction and $33.3\,\text{fs}$ FWHM pulse duration. The linearly polarized electric field (red-blue colored plot) oscillates in y-direction in the simulation plane (p-polarization). The laser intensity was chosen to $I = 3 \times 10^{18}\,\text{W/cm}^2$, hence $a_0 \approx 1.5$. The result is shown in figure 2.1.1, with the images ordered from top left to bottom right. The four images correspond to the begin of the interaction, the maximum of the laser field, the end of the pulse and the end of the simulation, respectively. The laser enters the simulation box from left. At some point in time the rising electric field starts to expel the electrons (black dots) out of the region of high intensity via the ponderomotive

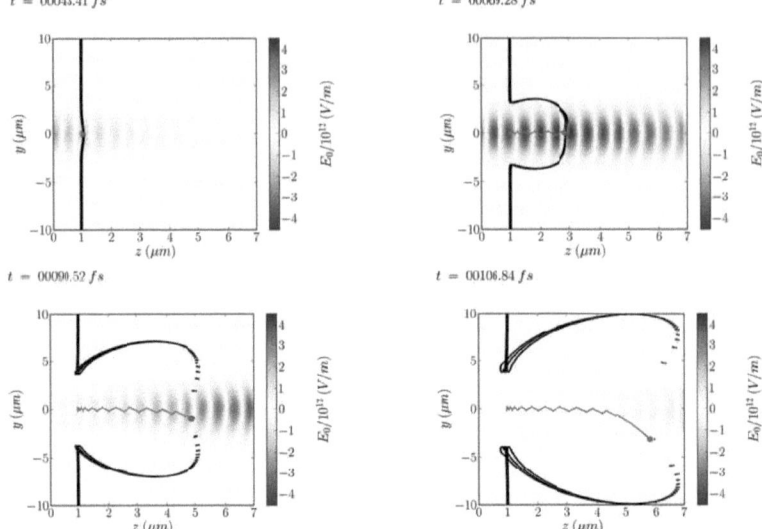

Fig. 2.2: 2D Particle-In-Cell (PIC) simulation of an intense laser interacting with free electrons. The laser pulse has $I = 3 \times 10^{18}\,\text{W/cm}^2$ and pulse duration of $33.3\,\text{fs}$. Images are ordered from top left to bottom right. The electric field E oscillates in y-direction (red-blue colored plot). At $z = 1\,\mu\text{m}$ free electrons were placed (black dots). The ponderomotive force of the laser expels the electrons in radial and forward direction. On axis a single electron is marked with a green line, it shows the oscillating trajectory as predicted by analytical theory.

force in forward and radial direction. Later on the electrons have been accelerated nearly completely out of the axis of symmetry. The green line marks the motion of a single electron (green dot) close to the axis of symmetry. There the interaction is nearly one-dimensional. The electron moves in a zig-zag motion in the electric field, as has been shown in fig. 2.1. At the end of the pulse the electron in the center has been decelerated again and finishes with nearly zero velocity as expected. The push downwards is due to the intensity gradient (ponderomotive force) since the electron is not exactly at the center. The ejection angle can be calculated to $\tan^2\theta = 2/(\bar\gamma - 1)$ [142–144].

2.2 Plasma interaction at the target front side

After this rather academic case of a laser interacting with free electrons in vacuum, the interaction with a solid is considered. The laser pulse in an experiment is not perfectly short-pulsed in time, but has a preceding pedestal or even pre-pulses in the ns to ps range with 10^{-4} to 10^{-7} times the intensity of the main pulse. Therefore already the onset of the high-intensity laser pulse with orders of magnitude less intensity is able to create a plasma at the surface of a solid, when the focused power exceeds the intensity of $I \approx 10^9 \, \text{W/cm}^2$ [145]. The electromagnetic wave couples onto free electrons, which oscillate in the laser electric field and ionize further atoms via inelastic collisions. Typical electron densities are about $n_i = 10^{21} \, \text{cm}^{-3}$ and ion densities are $n_i = n_e/Z_i$, where Z_i denotes the charge of the ions. A thin ablation plasma sheath is created at the surface that expands in vacuum with the ion sound speed $c_s = (k_B(Z_i T_e + T_i)/m_i)^{1/2}$, where k_B is the Boltzmann constant, T_e the electron temperature, T_i the ion temperature and m_i the ion mass. Under the assumption of a one-dimensional isothermal expansion [132] an exponentially decaying density profile develops with a scale length $l_s = c_s t$. The scale length for an exponentially decaying profile $n(z)$ is defined as the position where it is decayed to $1/\exp(1) = 0.368$ of the initial value and can be obtained by

$$l_s = \left[\frac{1}{n(z)} \frac{\mathrm{d}n(z)}{\mathrm{d}z}\right]^{-1}. \tag{2.14}$$

Typical scale lengths of plasma expansion before the main pulse arrives are on the order of a few micrometer. The plasma ablation leads to an inward-traveling shockwave due to momentum conservation, that compresses and heats the matter.

The plasma electrons are pushed by the laser which leads to an electric field due to the nearly immobile ion background, that forces the electrons to oscillate with the electron plasma frequency

$$\omega_p^2 = \frac{e^2 n_e}{\varepsilon_0 \bar{\gamma} m_e}. \tag{2.15}$$

With respect to the laser frequency ω_L the plasma is called *overdense* when $\omega_p > \omega_L$. This is the case beyond the critical density

$$n_c = \frac{\omega_L^2 \varepsilon_0 \bar{\gamma} m_e}{e^2}. \tag{2.16}$$

At this density, which can be calculated by $n_c = 1.1 \times 10^{21} \, \bar{\gamma}/\lambda_L \, [\mu m] \, \text{cm}^{-3}$, the

plasma refractive index
$$\eta = \sqrt{1 - \omega_p^2/\omega_L^2}, \qquad (2.17)$$

becomes imaginary and the laser wave can penetrate evanescently over a distance known as the collisionless skin depth $l_d = c/\omega_p$ only. For the relativistic case when $\bar{\gamma} > 1$, the critical density is higher than in the non-relativistic case due to the relativistically enhanced electron mass. Thus the laser light can even propagate further into the former overdense plasma, which is termed *relativistic transparency*. The relativistic interaction in the underdense part does not only increase the critical density, but also the plasma frequency decreases, which in turn leads to an intensity-dependent, thus spatially varying refractive index η. It is most strongly on axis, which acts analogous to a positive lens that *relativistically self-focuses* the beam even further.

2.2.1 Forward electron acceleration

Additionally, as it was shown in the section before, the ponderomotive force leads to a depletion of the electrons in the region of the laser pulse. This leads to density modulations in the wake of the laser pulse, moving with the group velocity $v_g = c\eta$. Electrons trapped in the electrostatic wake behind the laser can be efficiently *wakefield accelerated* [146]. For very short and intense pulses this scheme changes to the *bubble acceleration* scheme [147], resulting in mono-energetic electron beams [148–151] up to GeV energies [152–154].

While propagating in the plasma the laser can transfer energy to it via *inverse Bremsstrahlung* and *resonance absorption* [132], however they are of minor importance for intensities above $10^{18}\,\text{W/cm}^2$ [134]. When the density gradient in the preplasma is very strong, i.e., the scale length l_s is on the order of the laser wavelength, the phenomenon of *not-so-resonant resonance absorption* which is also known as *vacuum heating* or *Brunel effect* can occur [155]. In this case the laser drives an electrostatic wave at the critical density. The excursion of an electron in this wave is so strong that it is literally pulled out into the vacuum and sent back into the plasma in the next laser half-cycle. Since the laser cannot propagate beyond the critical density, this scheme of electron generation and heating can be quite effective.

Another important mechanism of laser heating is the *relativistic $\boldsymbol{j} \times \boldsymbol{B}$* heating [156], which is very similar to the vacuum heating but depends on the high-frequency $\boldsymbol{v} \times \boldsymbol{B}$-component of the Lorentz-force oscillating with twice the laser frequency, as can be seen by inserting eq. (2.3) in eq. (2.11). The heating caused by this high-frequency oscillation is analogous to the heating caused by a p-polarized electric field parallel to the density gradient. Hence it is most efficient for normal incidence and works for any polarization apart from circular. This was first confirmed by computer simulations [157] and later on experimentally by Malka and Miquel [158], who showed that for intensities above $I > 10^{19}\,\mathrm{W/cm^2}$ the electrons ejected along the laser axis direction can indeed be described by the relativistic $\boldsymbol{j} \times \boldsymbol{B}$ heating model. Since the heating can occur at a frequency of $2\omega_L$, twice every cycle a bunch of high-energy electrons will be generated, separated by a distance of half a laser wavelength, or $\approx \pi c/\omega_L$ [159–161]. For most high-intensity laser pulses $\lambda_L = 1\,\mu\mathrm{m}$, hence the electron bunches are created every 0.27 fs, accelerated along the gradient direction and are separated by $0.5\,\mu\mathrm{m}$. However, recent measurements [162] have shown that only a small fraction of less than 1 % of the laser energy is transferred to these micro-bunched electrons.

There are (too) many of several other absorption mechanisms, indicating that laser absorption and hot electron generation is still not well understood [163].

As pointed out by Bezzerides [164] the injection of the electrons into the laser wave is random, hence the energy gained by the electrons is randomly distributed around a central value, depending on the light pulse and plasma properties. The mean energy can be estimated by the ponderomotive potential, as will be shown below. The random injection results in a relativistic, three-dimensional particle density distribution of the electrons, which is known as Maxwell-Jüttner distribution [165]. The relativistic electrons are directed mainly in forward direction [166], hence the particle distribution function can be simplified by a one-dimensional Maxwell-Jüttner distribution, that is close to an ordinary Boltzmann distribution. A discussion on which distribution function best fits the experimental data is given in refs. [167, 168] and in more detail in ref. [169], leading to the conclusion that it is still not clear from neither theoretical nor experimental data to give a clear answer on the question about the shape of the distribution function. Therefore and for the sake of simplicity, it is approximated by

$$n_{\mathrm{hot}}(E) = n_0 \exp\left(-\frac{E}{k_B T_{\mathrm{hot}}}\right), \qquad (2.18)$$

that is determined both by the parameters $k_B T_{\mathrm{hot}}$ and n_0. The total number per unit volume n_0 can be estimated by the assumption that the amount of energy, per unit volume, transferred to the electrons via the ponderomotive force, is equal to the laser energy density:

$$n_0 \int f_{\text{pond}}\, \mathrm{d}^3 x \doteq \frac{E_L}{c\tau_L \pi r_0^2} \qquad (2.19)$$

where E_L is the laser energy, τ_L the laser pulse duration and r_0 the focal spot radius, respectively. The solution of the integral the mean energy of the electrons in the ponderomotive potential, which is the hot electron temperature $k_B T_{\text{hot}}$ that will be given below with eq. (2.25). In addition to that, the conversion efficiency from laser energy to hot electrons is not perfect, but only a fraction η is converted. Hence the total number of electrons generated by the laser pulse is given as:

$$n_0 = \frac{\eta E_L}{c\tau_L \pi r_0^2 k_B T_{\text{hot}}}. \qquad (2.20)$$

The fraction η was determined as being intensity-dependent as well, resulting in a scaling

$$\eta = 1.2 \times 10^{-15}\, I^{0.74}, \qquad (2.21)$$

where the intensity is given in W/cm² [15]. The scaling fits very well measured data in refs. [167,170] as well as simulations [157]. The maximum conversion efficiency was found to be $\eta_{\max} = 0.5$ [167, 170], however for ultrahigh intensity ($I > 10^{20}\,\text{W/cm}^2$) it can become up to 60 % for near-normal incidence and up to 90 % for irradiation under 45° [171].

As an example the number of electrons generated by an $I = 10^{19}\,\text{W/cm}^2$ laser pulse with $E_L = 20\,\text{J}$, $\tau_L = 600\,\text{fs}$, $\lambda_L = 1\,\mu\text{m}$ and focal spot radius $r_0 = 10\,\mu\text{m}$ results in

$$n_0 = 3.3 \times 10^{20}\,\text{cm}^{-3}. \qquad (2.22)$$

Hence, in the focal spot volume $V = \pi r_0^2 c \tau_L$ about

$$N = \frac{\eta E_L}{k_B T_{\text{hot}}} \approx 2 \times 10^{13} \qquad (2.23)$$

hot electrons will be generated. It is notable that the same relation (2.20) can be obtained in the simplistic picture of a non-relativistic ideal electron gas, that is compressed by the laser light. The ideal gas equation of state is $N k_B T_{\text{hot}} = pV$, where the pressure p is given by the light pressure $P_{\text{rad}} = I_L/c$ and $n_0 = N/V$. Solving for n_0 and inserting $I_L = E_L/(\tau_L \pi r_0^2)$ again leads to eq. (2.20).

The second parameter determining $n_{\text{hot}}(E)$ is the *hot electron temperature* $k_B T_{\text{hot}}$. For the exponential distribution of eq. (2.18) it corresponds to the mean energy, which can be estimated by the ponderomotive potential, eq. (2.13), to

$$k_B T_{\text{hot}} = m_o c^2 \left(\sqrt{1 + a_0^2/2} - 1 \right). \tag{2.24}$$

Although this is the correct derivation for a linearly polarized laser pulse, the literature is inconsistent since some authors use the formula from above [139, 140, 157, 158, 172], while others (e.g. [52, 157, 158, 173]) state the following expression:

$$k_B T_{\text{hot}} = m_o c^2 \left(\sqrt{1 + a_0^2} - 1 \right), \tag{2.25}$$

which is slightly different and corresponds to the ponderomotive potential for a circularly polarized laser pulse. Another notation for the hot-electron temperature is

$$k_B T_{\text{hot}} = m_o c^2 \sqrt{1 + \frac{2U_p}{m_o c^2}} \quad \text{with} \quad U_p = 9.33 \cdot 10^{-14} \frac{I}{\text{W/cm}^2} \frac{\lambda^2}{\mu \text{m}^2} \text{ eV}, \tag{2.26}$$

valid for the ultra-relativistic case ($a_0 \gg 0$), as given by refs. [52, 174].
A different hot electron temperature scaling sometimes used (e.g. [175, 176]) for hot electron generation in general, but originally determined for the *backward* electron generation, i.e., the antipodal laser direction out of the target, is

$$k_B T_{\text{front}} \approx 100 \, I_{17}^{1/3} \text{ [keV]}, \tag{2.27}$$

where I_{17} is the laser intensity in 10^{17} W/cm² [176].

The reason is, that on the one hand the ponderomotive potential is an *estimate* for the hot electron temperature only and on the other hand, for an evanescent or partially standing wave, which is the case at the critical density, no analytic expressions for W_{pond} are known [163]. In addition to that, in an experiment there is no single absorption and electron heating mechanism, but several effects contribute to the electron heating. These could be the density profile of the pre-plasma [177, 178] or its size [179], laser pre-pulse effects [180] as well as resonance absorption effects when the irradiation takes place under non-normal incidence [176], just to name a few. Another feature that becomes important for relatively long laser pulses on the order of 1 ps, is the radiation pressure $P_{\text{rad}} = 2I_L/c$. For intense laser pulses, e.g. for

$I_L = 5 \times 10^{19}\,\mathrm{W/cm^2}$, the radiation pressure $P_{\mathrm{rad}} = 3.3 \times 10^{15}\,\mathrm{Pa} = 3.3 \times 10^4\,\mathrm{MBar}$ is extremely high. The pressure pushes the critical surface inwards and successively drills a hole in the overdense plasma, hence the effect is called *laser-hole boring* [157, 181]. The hole boring is most effective in the center of the laser pulse, hence it leads to a convex deformation of the critical surface. The electric field then can couple better to the electrons at the sides of the hole, increasing the absorption and hot electron temperature [157, 182]. However, there is experimental evidence that ponderomotive acceleration is the major electron-acceleration mechanism, producing most of the fast electrons propagating in forward direction [161, 179].

Although it is clear that the hot electron temperature depends on I_0 and λ_L, the question remains which scaling best fits "reality". An extensive recherche in the literature, searching for data where the hot electron temperature was either measured or obtained via computer simulations, has lead to the plot shown in figure 2.3.

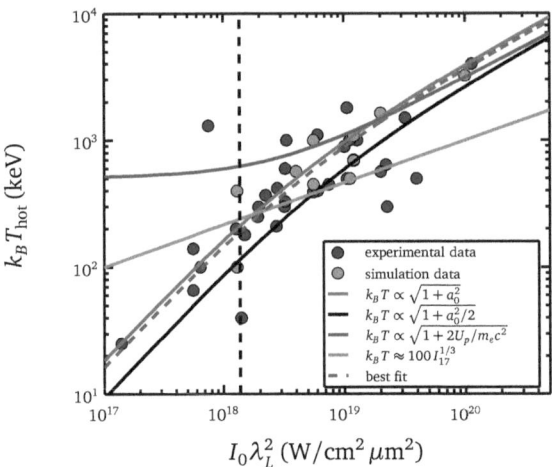

Fig. 2.3: Plot of the hot electron temperature $k_B T_{\mathrm{hot}}$ versus laser intensity $I\lambda^2$, obtained by measurements (•) or computer simulation (∘). The data are compared to various scaling laws, explained in the text. The dashed, vertical line represents the threshold of relativistic interaction where $a_0 = 1$.

The data was extracted from refs. [58, 157, 162, 167, 172, 173, 175, 182–185] as well as ref. [133], p. 178. It resembles a large variety of intensities $I = [10^{17} - 10^{20}]\,\mathrm{W/cm^2}$, of wavelengths $\lambda_L = [0.248 - 1.064]\,\mu\mathrm{m}$, of irradiation angles from $0°$ up to $45°$ and of s- and p-polarized incidence. The contrast ratio of the pre-pulse level to main pulse was stated being 10^{-6} or better. The blue circles represent measured data, the green circles correspond to data obtained by computer simulations. For a clearer picture, and since it was not always given, there are no error bars plotted. Although the figure axes are bi-log plotted, the data scatter relatively large. However, a trend of increasing $k_B T_{\text{hot}}$ with increasing $I_0 \lambda_L^2$ is visible. The black line corresponds to a plot of eq. (2.24), the red line shows the plot of eq. (2.25). The brown line shows the scaling for the ultra-relativistic case from eq. (2.26). The grey line, with a different scaling, is obtained from eq. (2.27). The intensity threshold, where relativistic interaction begins, is depicted with the dashed, vertical line.

There is a remarkable agreement between the majority of data points and the scaling for the ponderomotive potential for a circularly polarized wave from eq. (2.25). A best fit to all data points reveals a scaling of $k_B T_{\text{hot}} = m_e c^2 (1 + a_0^2 / 1.14)^{1/2}$ (magenta line), very close to the ponderomotive scaling. Hence it is legitimate to assume that

$$k_B T_{\text{hot}} = m_0 c^2 \left(\sqrt{1 + \frac{I_0 \, [\text{W/cm}^2] \lambda_L^2 \, [\mu\text{m}^2]}{1.37 \times 10^{18}}} - 1 \right), \qquad (2.28)$$

is an adequate assumption for the hot electron temperature scaling in relativistic laser-plasma interaction with solid targets. As an example, using the same laser parameters as before, with $I_0 \lambda_L^2 = 10^{19}\,\text{W/cm}^2$, a hot electron temperature of about 1 MeV can be anticipated.

2.3 Laser-ion acceleration

An intense laser pulse impinging onto a solid target is able to create MeV-electrons as shown in the sections before. Although the laser pulse is very intense, a *direct* laser-ion acceleration is strictly speaking not happening. The quiver motion of protons, that are the lightest ions, in the laser pulses scales with $v_{\text{osc}} \propto m_p^{-1}$ (eq. (2.6)), which is about a factor of 2000 less than the electron motion. A forward acceleration, which requires the laser's magnetic field and hence $v_{\text{osc}} \approx c$ needs an intensity of $I \approx 5 \times 10^{24}\,\text{W/cm}^2$ according to eq. (2.6), where the electron mass has to be replaced by the ion mass.

However, the laser pushes the electrons, which, in turn, then interact with the remaining ions via the electric force due to charge-separation. An advantage in that is, that the fields created in this manner can be as high as the laser field itself, but since the ion motion is about a factor of $(m_i/Z_i m_e)^{1/2}$ slower, they remain stationary for relatively long times compared to the laser oscillation. Hence they are called *quasi-static*. The ions can gain energy by the potential of this field, which is just the ponderomotive potential. The ions start to blow-off and can gain energy on the order of a few MeV [51]. In laser-matter interaction a variety of ion-acceleration schemes were identified, e.g. long-pulse (nanosecond) plasma thermal expansion [51, 186], Coulomb explosion of laser-irradiated clusters [187], transverse acceleration in underdense plasma channels [188], ion acceleration in a charge-separation field by a quasi-stationary magnetic field [189] or acceleration from the shock front induced by laser hole boring [190]. Low-energy, but high-current ion beams can be produced by the skin-layer ponderomotive acceleration [191, 192] with sub-relativistic intensities. Theoretical studies have identified a very efficient acceleration of ultra-thin (nm-sized) foils by circularly polarized, ultra-high contrast laser radiation [123]. For linear polarization and ultra-high contrast as well, the irradiation of a nm-thin foil can lead to GeV energies by the laser-breakout afterburner effect [193]. Future generations of high-energy, high-intensity lasers with $I_L > 1.37 \times 10^{23}\,\mathrm{W/cm^2}$ and ultra-high contrast might be able to enter the laser-piston acceleration regime. There the radiation pressure can directly accelerate ions to GeV energies [194].

The scope of this thesis is on the ion-acceleration process that seems to be most efficient in terms of beam quality and usability to date: the acceleration of ions from the rear, i.e., the non-irradiated sides of thin (thickness $d = [5 - 50]\,\mu\mathrm{m}$) solid foil targets. It was discovered in 1999 by groups at Rutherford Appleton Laboratory, UK [18], at Lawrence Livermore National Laboratory (LLNL), USA [21, 22] as well as University of Michigan, USA [19]. Shortly after that, the mechanism was quantitatively explained by Wilks *et al.* [52] and is known since then as the *target normal sheath acceleration – TNSA*.

A schematic of the acceleration process is shown in figure 2.4. The preceding pedestal of the laser (the pre-pulse) with an intensity on the order of 10^{-6}-fold the main pulse intensity creates a plasma at the target's front side. Pre-pulses can be created by amplified spontaneous emission (ASE) as well as spectral modulations in the process of chirped-pulse amplification and compression. The main pulse then interacts with the plasma, accelerating electrons that are directed mainly in forward direction as discussed in the section before. The electrons are transported through the foil, where collisions with the background material could increase the initial divergence of the electron current. Issues of the electron transport are given in the next sub-

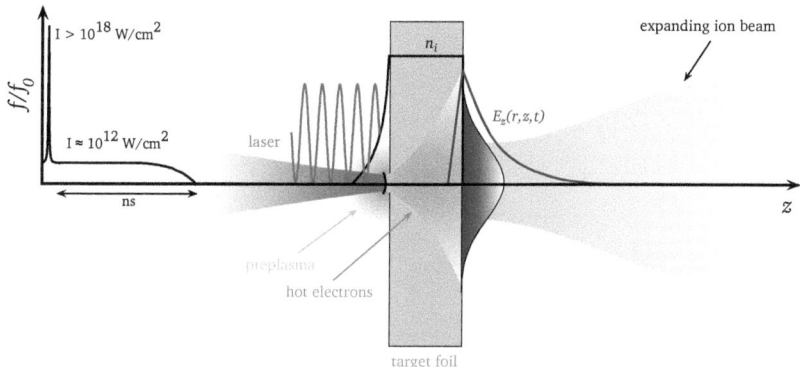

Fig. 2.4: Target Normal Sheath Acceleration – TNSA. A thin target foil with thickness $d = [5-50]\,\mu$m is irradiated by an intense laser pulse. The laser pre-pulse creates a pre-plasma on the target's front side. The main pulse interacts with the plasma and accelerates MeV-energy electrons mainly in forward direction. The electrons propagate through the target, where collisions with the background material can increase the divergence of the electron current. The electrons leave the rear side, resulting in a dense sheath. An electric field due to charge-separation is created. The field is on the order of the laser's electric field (TV/m), which ionizes atoms at the surface. The ions are then accelerated in this sheath field, pointing in the target normal direction.

section. The electrons then leave the rear side, forming a dense electron cloud. The charge-separation of the electrons from the remaining target creates a strong electric field on the order of TV/m in a thin sheath. The field ionizes atoms at the rear side, e.g. protons and carbon ions from contamination layers [51]. The ions are then accelerated along the target normal direction by this field, gaining energies up to tens of MeV. This TNSA mechanism is further discussed in subsection 2.3.2.

2.3.1 Fast-electron transport in dense matter

The transport of fast electrons in dense matter is still a very active research field, because the experimental as well as the theoretical access are both quite complex due to the high current, high density and the non-linear interaction involved. An overview of the field can be found in ref. [168].

A schematic picture of the electron transport in intense laser-matter interaction is shown in fig. 2.5 and will be used for the explanation of the various effects observed in electron transport. The laser impinges from the left side on a pre-formed, exponentially decaying pre-plasma (sec. 2.2). The light pressure pushes the critical surface n_c, leading to a steeping of the density profile. The plasma ablation drives

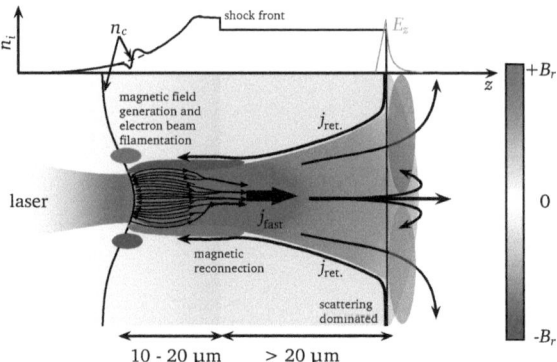

Fig. 2.5: Schematic of laser-generated fast-electron transport. The laser (shown in red) impinges on a pre-plasma with exponential density profile from the left side. The light pressure leads to profile steepening, depicted in the one-dimensional scheme on top. An ablation plasma creates an inward-traveling shockwave, that heats, ionizes and compresses the target. Fast electrons are created by the laser, propagating into the dense plasma towards the target rear side. The high electron current j_{fast} can lead to filamentation and magnetic field generation (shown by the light red- and blue-colored areas), as well as it drives a return current $j_{\text{ret.}}$. The global magnetic field tends to pinch the fast-electron current. Electrons propagating in the dense, solid matter interact by binary collisions with the background material. This leads to a broadening, that becomes the major effect for longer distances. At the rear side, the electrons form a sheath and build up an electrostatic field E_z (grey line in 1D-plot). This can lead to re-fluxing (re-circulation) of the electrons, heating the target even more.

an inward-traveling shock wave, leading to ionization as well as a temperature increase in the former cold solid.

The laser creates a hot electron distribution (sec. 2.2.1), which is accelerated into the dense plasma. The estimate in section 2.2.1 has shown, that about $N = \eta E_L / k_B T_{\text{hot}}$ electrons with energies in the MeV-range are created in intense laser-matter interaction. They are injected into the dense plasma with an angular distribution according to $\tan\theta = [2/(\gamma - 1)]^{1/2}$ [173, 195]. The injection direction depends on the direction of the pre-plasma density gradient as well as the laser beam propagation direction [179].

The huge number of electrons amounts to a current of $j_{\text{fast}} = e N/\tau_L$, that is on the order of mega-ampere for typical laser parameters. Hence the propagation of fast electrons is not only governed by collisional effects, determining the stopping power of electrons in the material, but collective (e.g. electromagnetic) effects as well. Assuming a straight electron transport [196] in a cylinder with radius of the laser spot radius, the current creates a magnetic flux density $B = \mu_0 j_{\text{fast}}/2\pi r_0$ on the order of 10^5 T. The magnetic field energy is then

$$W = \frac{\mu_0 e^2}{8\pi} \left(\frac{\eta E_L}{\tau_L k_B T_{\text{hot}}} \right)^2 d, \qquad (2.29)$$

where d denotes the target thickness, that is $d = [5 - 50]\,\mu$m typically. The energy stored in this field easily becomes greater than the laser energy, which violates energy conservation. For typical experimental parameters, e.g. taking the example from sec. 2.2.1, the limit is reached for $d \approx 10\,\mu$m already. Hence for a transport from the front to the rear side a return current $j_{\text{ret.}}$ must exist, balancing the forward-directed current to yield $\boldsymbol{j}_{\text{total}} = \boldsymbol{j}_{\text{fast}} + \boldsymbol{j}_{\text{ret.}} \approx 0$. In addition to that, without the return current the electric field according to $\partial E/\partial t = -j_{\text{fast}}/\varepsilon_0$ would stop the electrons in a distance of less than 1 nm [197]. The electric field driving the return current in turn, can be strong enough to stop the fast electrons. The effect is known as *transport inhibition*, being significant in insulators [198–200], but negligible in conductors (to first order) [199]. Both magnetic field as well as electric field generation are inversely dependent on the target's electrical conductivity, hence conducting targets are favorable for laser-ion acceleration, where a transport from the front- to the rear-side is necessary [76]. The condition $\boldsymbol{j}_{\text{total}} = \boldsymbol{j}_{\text{fast}} + \boldsymbol{j}_{\text{ret.}} \approx 0$ implies that the number of fast electrons is much smaller than the slow electrons carrying the return current [197]. The counter-propagating streams of charged particles are subject to a nonlinear Weibel instability [201], leading to a filamentation of the electron beams [202,203] in the low-density part of the target (pre-plasma) and self-generated magnetic field filamentation. The gyroradius of the electrons is on the order of the local skin length $l_s = c/\omega_p \approx 0.1\,\mu$m [204]. The large magnetic fields accompanying the laser-plasma interaction are depicted by the red and blue-colored areas in fig. 2.5. The red color depicts a field going out of the drawing plane, the blue color represents an inward-going field. These strong fields were seen in computer simulations [204–208] and evidence was found in proton-radiography experiments [25,209]. Due to these filaments a much higher current can be transported, which is well above the Alfvén current $I_{\text{Alfvén}} \approx 17\,\beta\gamma$ kA [210, 211]. The filaments can coalesce due to mutual magnetic attraction over longer spatial scales [207], and can channel to a single magnetized jet later on [212]. In addition to that there is experimental evidence for a micrometer-sized, filamented [213, 214], possibly jet-like [215] or hollow electron transport in insulators [216] and layered targets [76, 217], but not in metal foils where a smooth and uniform transport was found [76, 213]. The reason for a filamentation in insulators is the so-called *ionization instability* [218] at the ionization front of the electron cloud propagation.

Since there is a global net electron current towards the rear side it is accompanied by a global magnetic field. With growing plasma density this field tends to pinch the electron current and it could lead to a collimated guiding of the whole electron distribution towards the target's rear side [219–222].

As soon as the electrons penetrate the cold solid region, binary collisions (multiple small-angle scattering) with the background material are no longer negligible. These tend to broaden the electron distribution, counteracting the magnetic field effect [223]. For long propagation distances ($z \gtrsim 15\,\mu$m), the current density is low enough, so that broadening due to small-angle scattering becomes the dominating mechanism [162].

When the electrons reach the rear side, they form a dense charge-separation sheath. The out-flowing electrons lead to a toroidal magnetic field $\boldsymbol{B_\theta}$, that can spread the electrons over large transverse distances by a purely kinematic $\boldsymbol{E} \times \boldsymbol{B_\theta}$-force [36], sometimes called *fountain effect* [31]. The electric field created by the electron sheath is sufficiently strong to deflect lower-energy electrons back into the target, which then *re-circulate*. Experimental evidence for recirculating electrons was found in refs. [172, 175, 198, 200]. Its relevance to proton acceleration was first demonstrated by Mackinnon *et al.* [65], who measured a strong enhancement of the maximum proton energy for thin foils below $10\,\mu$m, compared to thicker ones. With the help of computer simulations this energy-enhancement was attributed to an enhanced sheath density due to re-fluxing electrons. Further evidence of re-fluxing electrons was found in an experiment during this thesis, discussed in section 5.1 and published in ref. [2].

The majority of data shows a divergent electron transport. Measurements were performed with time-integrated imaging K_α-spectroscopy of buried tracer layers [213], time-resolved optical diagnostics of optical self-emission (transition radiation) at the target rear side [175], shadowgraphy and XUV-imaging of the rear side [224] as well as proton emission [76]. The transport full-cone angle of the electron distribution was determined to be dependent on laser energy, intensity as well as target thickness [182, 224]. For rather thick targets ($d > 40\,\mu$m) this value is around 30° FWHM [76, 175, 213] at the laser intensities used in this thesis, whereas for thin targets ($d \leq 10\,\mu$m) published values are in the range of 16° (indirectly obtained by a fit to proton energy measurements [17, 59]) and $\approx 150°$ at most [213, 224]. Just recently it was shown that different diagnostics lead to different electron transport cone angles [224], so the question about the 'true' cone angle dependence with laser and target parameters still remains unclear.

Further complications can arise by the observation that the laser focal spot shape imprints in the electron-transport pattern to the rear side. This was first observed by Fuchs et al. [76] and was further investigated in experiments during this thesis. The results with respect to the electron transport angle are in broad agreement to the references cited above. Details are given in section 5.1 and in ref. [2].

Neglecting the complicated interaction for thicknesses below $d \approx 15\,\mu$m, a reasonable estimate for the electron beam divergence is the assumption, that the electrons are generated in a region of the size of the laser focus and are purely collisionally transported to the rear side. This is in agreement with most published data. The broadening of the distribution is then due to multiple Coulomb small-angle scattering, given analytically e.g. by Moliére's theory in Bethe's description [225]. An extensive review on multiple small-angle scattering of charged particles in matter is given in [226].

To lowest order the angular broadening $f(\theta)$ follows a Gaussian (see ref. [225], eq. (27))

$$f(\theta) = \frac{2\,e^{-\vartheta^2}}{\chi_c^2 B} \sqrt{\theta/\sin\theta}, \tag{2.30}$$

where the second term on the right-hand side is a correction for larger angles (from [225], eq. (58)). The angle ϑ can be related to θ by $\vartheta = \theta/\chi_c B^{1/2}$. The transcendental equation $B - \ln B = \ln\left(\chi_c^2/\chi_{a'}^2\right)$ determines B. The screening angle $\chi_{a'}^2$ is given by $\chi_{a'}^2 = 1.167\,(1.13 + 3.76\alpha^2)\,\lambda^2/a^2$, where $\lambda = \hbar/p$ is the deBroglie wavelength of the electron and $a = 0.885\,a_B Z^{-1/3}$, with the Bohr radius a_B (see start of chapter 2 for the definition of a_B). α is determined by $\alpha = Ze^2/(4\pi\varepsilon_0 \hbar \beta c)$ with the nuclear charge Z, electron charge e, $\beta = v/c$ and $\varepsilon_0, \hbar, c$ denote the usual constants. The variable χ_c is given by

$$\chi_c^2 = \frac{e^4}{4\pi\varepsilon_0^2 c^2} \frac{Z(Z+1)\,N\,d}{\beta^2\,p^2}, \tag{2.31}$$

with the electron momentum p and $N = N_A \rho/A$ being the number of scattering atoms, determined by Avogadro's number N_A, material density ρ and mass number A. χ_c is proportional to the material thickness d and density ρ as $\chi_c \propto (\rho\,d)^{1/2}$. Since χ_c determines the width of $f(\theta)$, the angular spread of the electron distribution propagating through matter is proportional to its thickness as well as its density.

The analytical formula allows to estimate the broadening of the laser-accelerated electron distribution during the transport through the cold solid target. For a laser intensity $I_L = 10^{19}\,\text{W/cm}^2$ the mean energy (temperature) is $k_B T_\text{hot} \approx 1\,\text{MeV}$. The increase of the radius r with target thickness d is shown in figure 2.6. The electrons were chosen to propagate in Aluminum (———) and Gold (– – –). Al does not lead to a strong broadening due to its low density and Z, compared to the broadening in gold. The graph shows that in both cases the radius at the rear side scales as $r \propto d^2$ (———).

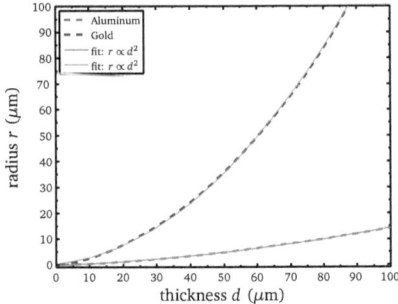

Fig. 2.6: Increase of the radius r of an electron distribution with target thickness d. The calculation was done with eq. (2.30), taking an energy of $k_B T_{\text{hot}} \approx 1\,\text{MeV}$, corresponding to a laser intensity $I = 10^{19}\,\text{W/cm}^2$. (– – –) shows the calculation for Gold, (———) corresponds to Aluminum. Both curves resemble a quadratic increase with thickness d (———).

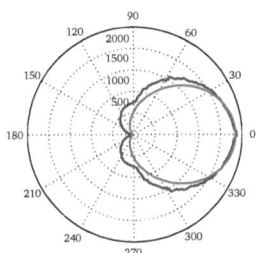

Fig. 2.7: Polar plot of a 3D electron distribution with 10 MeV temperature, see text for details.

The simple analytical formula for the calculation of the electron distribution broadening from above will now be compared to a more sophisticated calculation, performed by M. Günther with the GEANT4-code package [227], developed at CERN. The Monte-Carlo code includes a wealth of physics models to calculate the interaction and propagation of charged particles in matter. It was used to calculate the electron transport in a thick (6 mm) solid gold piece for nuclear activation studies [228], where the data plotted here are taken from. The input is an electron distribution of a 3D-Maxwell-Jüttner-type with 10 MeV temperature. Figure 2.3.1 shows a polar plot of the angle distribution $f(\theta)$ of the whole electron ensemble calculated by GEANT4 (———). The electrons mainly propagate in forward direction (0°), due to their relativistic velocity. Eq. (2.30) was used to calculate the broadening of electrons with energy $E = 10\,\text{MeV}$, that is just the average energy in

GEANT4. The FWHM of the resulting Gaussian distribution (——) is 122°, representing a mean value of the whole electron distribution broadening. Both curves are in arbitrary units and were normalized to each other. Although GEANT4 includes secondary electron generation and shower processes, both distributions are in reasonable agreement even for a very thick target, showing that the broadening of the whole ensemble can indeed be estimated by the broadening of electrons with an energy corresponding to the temperature.

The estimate based on an electron distribution broadening determined by small-angle scattering will be used in sec. 5.1.1 for an explanation of the measured proton beam profiles. It should be noted, that even though the model seems to be able to calculate the broadening of the forward-propagating fast electron distribution generated by intense laser-matter interaction, it could fail to determine the real number of electrons arriving at the rear side. According to Davies [169] the generation of electromagnetic fields as well as recirculation of the electrons have to be taken into account, both making an estimate and even calculation very difficult. Recent experiments by Akli *et al.* [229] have shown that this is true at least for thin targets below 20 μm, but for thicker foils the assumption of strong recirculation overestimates the number of electrons. Therefore the question if electromagnetic fields and recirculation are essential to determine the fast-electron transport from the front to the rear side can still not be satisfactorily answered, though making the assumption of simple collisional broadening a relatively good estimate.

2.3.2 Target Normal Sheath Acceleration - TNSA

After the transport through the target, the electrons end up at the rear side. The laser creates about 10^{13} electrons that are potentially all propagating through the target. The broadening results in transverse extension, that can be estimated by

$$r_{\text{sheath}} = r_0 + d\tan(\theta/2), \tag{2.32}$$

where r_0 denotes the laser spot radius, d the target thickness and θ the broadening angle of the distribution, e.g. calculated by eq. (2.30). The electrons exhibit an exponential energy distribution (eq. (2.18)) with temperature $k_B T$ and overall density n_0 given by eq. (2.20). The electron density at the rear side (neglecting recirculation) therefore can be estimated to

$$n_{e,0} = \frac{\eta E_L}{c \tau_L \pi (r_0 + d\tan\theta/2)^2 k_B T_{\text{hot}}} \tag{2.33}$$

$$\approx 1.5 \times 10^{19} \frac{r_0^2}{(r_0 + d\tan\theta/2)^2} \frac{I_{18}^{7/4}}{\sqrt{1 + 0.73 I_{18}\lambda_{\mu m}^2} - 1} \text{ cm}^{-3}. \tag{2.34}$$

The last equation was obtained by inserting equations (2.1), (2.20), (2.21) and (2.28) in the first one. I_{18} means that the intensity has to be taken in units of 10^{18} W/cm^2. The estimate shows that the electron density at the rear side strongly scales with the laser intensity and is inversely proportional to the square of the target thickness. Taking the standard example of a laser pulse with $I = 10^{19}$ W/cm^2, focused to a spot of $r_0 = 10\,\mu m$ and assuming a target thickness $d = 20\,\mu m$, the angular broadening according to eq. (2.30) is $\theta = 42°$ (FWHM) for electrons with mean energy $k_B T$, determined by eq. (2.28). Hence the electron density at the target's rear side is $n_{e,0} = 1.4 \times 10^{20}$ cm^{-3}. This is orders of magnitude below solid density and justifies the assumption of a shielded transport through the target.

The electrons arrive at the rear side and escape into vacuum. The charge separation leads to an electric potential Φ in the vacuum region, according to Poisson's equation. In a one-dimensional consideration it is given as

$$\varepsilon_0 \frac{\partial^2 \Phi}{\partial z^2} = e\, n_e, \tag{2.35}$$

For a solution it is assumed that the solid matter in one half-space ($z \leq 0$) perfectly compensates the electric potential, whereas for $z \to \infty$ the potential goes to infinity. Its derivative $\partial \Phi / \partial z$ vanishes for $z \to \pm\infty$.

In the vacuum region ($z > 0$), the field can be obtained analytically [230]. The electron density is taken as

$$n_e = n_{e,0} \exp\left(\frac{e\Phi}{k_B T_{\text{hot}}}\right), \qquad (2.36)$$

where the electron kinetic energy is replaced by the potential energy $-e\Phi$. The initial electron density $n_{e,0}$ is taken from eq. (2.34). The solution of the Poisson equation is found with the Ansatz $e\Phi/k_B T_{\text{hot}} = -2\ln(\lambda z + 1)$, where λ is a constant defined by the solution and the $+1$ is necessary to fulfill a continuos solution with the condition $\Phi(0) = 0$ at the boundary to the solid matter. The resulting potential is

$$\Phi(z) = -\frac{2k_B T_{\text{hot}}}{e} \ln\left(1 + \frac{z}{\sqrt{2}\lambda_D}\right) \qquad (2.37)$$

and the corresponding electric field reads

$$E(z) = \frac{2k_B T_{\text{hot}}}{e} \frac{1}{z + \sqrt{2}\lambda_D}. \qquad (2.38)$$

In this solution the electron Debye length

$$\lambda_D = \left(\frac{\varepsilon_0 k_B T_{\text{hot}}}{e^2 n_{e,0}}\right)^{1/2} \qquad (2.39)$$

appears, that is defined as the distance over which significant charge separation occurs [231]. Replacing $k_B T_{\text{hot}}$ with eq. (2.28) and $n_{e,0}$ with eq. (2.34) leads to

$$\lambda_D \approx 1.37\,\mu\text{m}\, \frac{r_0 + d\tan\theta/2}{r_0} \frac{\sqrt{1 + 0.73\,I_{18}\lambda^2} - 1}{I_{18}^{7/8}}. \qquad (2.40)$$

The Debye length, or longitudinal sheath extension, on the rear side is on the order of a micrometer. It scales quadratically with target thickness (since $d\tan(\theta/2) \propto d^2$, see scaling in fig. 2.6) and is inversely proportional to the laser intensity. Thus, a higher laser intensity on the front side leads to a shorter Debye length at the rear side and results in a stronger electric field. The standard example from above leads to $\lambda_D = 0.6\,\mu\text{m}$.

The maximum electric field is obtained at $z = 0$ to

$$E_{max}(z=0) = \frac{\sqrt{2}k_B T_{\text{hot}}}{e\lambda_D} \tag{2.41}$$

$$\approx 5.2 \times 10^{11}\,\text{V/m}\,\frac{r_0}{r_0 + d\tan\theta/2}\,I_{18}^{7/8} \tag{2.42}$$

$$= 9 \times 10^{10}\,\text{V/m}\,\frac{r_0}{r_0 + d\tan\theta/2}\,E_{12}\,E_{12}^{3/4}. \tag{2.43}$$

Hence the initial field at $z = 0$ is proportional to the laser intensity and it depends nearly quadratically on the laser's electric field strength. In the last equation the laser's electric field strength is inserted in normalized units of 10^{12} V/m. By inserting the dependence of the broadening with target thickness from fig. 2.6, the scaling with the target thickness is obtained as $E_{max}(z=0) \propto d^{-2}$. The standard example leads to a maximum field strength of $E_{\max} \approx 2 \times 10^{12}$ V/m just at the surface, that is on the order of TV/m or MV/μm. It is only slightly less than the laser electric field strength of $E_0 = 8.7 \times 10^{12}$ V/m. However, for later times than $t = 0$ the field strength is dictated by the dynamics at the rear side, e.g. ionization and ion acceleration.

As just mentioned, the electric field strength instantly leads to ionization of the atoms at the target rear surface, since it is orders of magnitude above the ionization threshold of the atoms. A simple model to estimate the electric field strength necessary for ionization is the *Field Ionization by Barrier Suppression (FIBS)* model [232]. The external electric field of the laser overlaps with the Coulomb potential of the atom and deforms it. As soon as deformation is below the binding energy of the electron, it is instantly freed, hence the atom is ionized. The threshold electric field strength E_{ion} can be obtained with the binding energy U_{bind} as

$$E_{\text{ion}} = \frac{\pi\varepsilon_0 U_{\text{bind}}^2}{e^3 Z}. \tag{2.44}$$

As the electron sheath at the rear side is relatively dense, the atoms could also be ionized by collisional ionization. However, as discussed by Hegelich [49] the cross section for field ionization is much higher than the cross section for collisional ionization for the electron densities and electric fields appearing at the target surface. Taking the ionization energy of an hydrogen atom with $U_{\text{bind}} = 13.6$ eV, the field strength necessary for FIBS is $E_{\text{ion}} = 10^{10}$ V/m. It is two orders of magnitude below the field strength developed by the electron sheath in vacuum as shown above. Hence nearly all atoms (Protons, Carbons, heavier particles) at the rear side are instantly ionized and, since they are no longer neutral particles, they are then subject to the electric field and are accelerated. The maximum charge state of ions found in an experiment is an estimate of the maximum field strength that appeared. This has been used to estimate the sheath peak electric field value [49] as well as the field extension in transverse direction [63, 75].

The strong field ionizes the target and accelerates ions to MeV-energies, if it is applied for long enough time. The time can be easily calculated by the assumption of a test-particle moving in a static field, generated by the electrons. Free protons were chosen as test-particles. The non-linear equation of motion is obtained from eq. (2.38). The solution was obtained numerically with MATLAB [233]. It shows that for a proton to obtain a kinetic energy of 5 MeV, the field has to stay for 500 fs in the shape given by eq. (2.38). During this time the proton has travelled $11.3\,\mu$m. The electric field will be created as soon as electrons leave the rear side. Some electrons can escape this field, whereas others with lower energy will be stopped and will be re-accelerated back into the target. Since the electron velocity is close to the speed of light and the distances are on the order of a micrometer, this happens on a few-fs time scale, leading to a situation where electrons are always present outside the rear side. The electric field being created does not oscillate but is quasi-static on the order of the ion-acceleration time. Therefore ultra-short laser pulses, although providing highest intensities, are not the optimum laser pulses for ion acceleration. The electric field is directed normal to the target rear surface, hence the direction of the ion acceleration follows the target normal, giving the process its name *Target Normal Sheath Acceleration – TNSA* [22, 52].

2.4 Expansion models

The laser-acceleration of ions from solid targets is a complicated, multi-dimensional mechanism including relativistic effects, non-linearities, collective as well as kinetic effects. Theoretical methods for the various physical mechanisms involved in TNSA range from analytical approaches for simplified scenarios over fluid models up to fully relativistic, collisional three-dimensional computer simulations.

Most of the approaches that describe TNSA neglect the complex laser-matter interaction at the front-side as well as the electron transport through the foil. These *plasma expansion models* start with a hot electron distribution that drives the expansion of an initially given ion distribution [15, 17, 61, 67–70, 101, 230, 234–239]. Crucial features like the maximum ion energy as well as the particle spectrum can be obtained analytically, whereas the dynamics have to be obtained numerically. The plasma expansion description dates back to 1954 [240]. Since then various refinements of the models were obtained, with an increasing activity after the first discovery of TNSA. These calculations resemble the general features of TNSA. Nevertheless, they rely on somewhat idealized initial conditions from simple estimates. In addition to that, the plasma expansion models are *one-dimensional*, whereas the experiments have clearly shown that TNSA is at least two-dimensional. Hence these models can only reproduce one-dimensional features, e.g., the particle spectrum of the TNSA process.

Sophisticated three-dimensional computer simulation techniques have been developed for a better understanding of the whole process of short-pulse high-intensity laser-matter interaction, electron transport and subsequent ion acceleration. The simulation methods can be classified as (i) Particle-In-Cell (PIC), (ii) Vlasov, (iii) Vlasov-Fokker-Planck, (iv) hybrid fluid/particle and (v) gridless particle codes; see the short review in ref. [197] for a description of each method. The PIC method is the most widely used simulation technique, and it was used in the framework of this thesis, too. In PIC the Maxwell equations are solved, together with a description of the particle distribution functions. The method resembles more or less a "numerical experiment" with only little approximations, hence a detailed insight into the dynamics can be obtained. The disadvantage is that no specific theory serves as an input parameter and the results have to be analyzed like experimental results, i.e., they need to be interpreted and compared to analytical estimates.

The next section describes the two models - the plasma expansion model and a PIC code - in more detail. Both methods were used for an explanation of the experimental results presented in chapter 4 and lead to the development of a three-dimensional particle transfer code that is able to fully reproduce a measured proton beam. This development will be presented in chapter 5.

2.4.1 Plasma expansion model

The plasma expansion models of TNSA are close to the expansion of a isothermal rarefaction wave in a freely expanding plasma [132, 241]. The model makes the assumption of *quasi-neutrality* $n_e = Z n_i$ and of constant temperature T_e. This isothermal expansion is described by a two-fluid hydrodynamic model of electrons and ions. The governing equations are the

continuity equations:
$$\frac{\partial n_{e,i}}{\partial t} + \nabla \cdot (n_{e,i}\boldsymbol{v}_{e,i}) = 0, \qquad (2.45)$$

momentum conservation:
$$m_e n_e \frac{d\boldsymbol{v}_e}{dt} = -n_e e (\boldsymbol{E} + \boldsymbol{v}_e \times \boldsymbol{B}) - \nabla p_e + \boldsymbol{f}_e,$$
$$m_i n_i \frac{d\boldsymbol{v}_i}{dt} = -n_i Z e (\boldsymbol{E} + \boldsymbol{v}_i \times \boldsymbol{B}) - \nabla p_i + \boldsymbol{f}_i, \qquad (2.46)$$

and energy conservation:
$$\frac{3}{2} m_e k_B \frac{dT_e}{dt} = -p_e \nabla \boldsymbol{v}_e + h_e - \nabla \boldsymbol{q}_e,$$
$$\frac{3}{2} m_i k_B \frac{dT_i}{dt} = -p_i \nabla \boldsymbol{v}_i + h_i - \nabla \boldsymbol{q}_i. \qquad (2.47)$$

The electron and ion densities are denoted by n_e and n_i, the electron and ion velocities are represented by v_e and v_i, respectively. The ∇-operator denotes the derivative with respect to (x, y, z). The electric field is $\boldsymbol{E}$ and $\boldsymbol{B}$ denotes the magnetic field. The electron and ion pressure tensors are p_e and p_i, whereas $\boldsymbol{f}_{e,i}$ denote any external forces. The temperature of the electron and ion fluids are represented by $k_B T_{e,i}$. Any possible externally applied heat is marked by h_e, h_i. $\boldsymbol{q}_e$, $\boldsymbol{q}_i$ describe heat conduction. The total derivative is taken as $d/dt = \partial/\partial t + \boldsymbol{v}_e \nabla$.
Next, the following simplifications are assumed:

- isothermal expansion: $\frac{dT}{dt} = 0$

- no heating (no laser): $h_{i,j} = 0$

- no heat conduction: $\nabla \boldsymbol{q} = 0$

- no collisions, that is no viscosity and no external forces: $\boldsymbol{f} = 0$

- electrostatic acceleration: no magnetic fields, $\boldsymbol{B} = 0$

Therefore energy conservation implies $\nabla \boldsymbol{v}_{i,e} = 0$, which means an incompressible fluid. The three conservation laws provide 10 equations to solve 13 variables $(n_e, n_i, \boldsymbol{v}_e, \boldsymbol{v}_i, \boldsymbol{E}, p_e, p_i)$. The three missing relations are the quasi-neutrality condition $Zn_i = n_e$ and the material equations, namely the ideal gas equations:

$$p_e = n_e k_B T_e \qquad\qquad p_i = n_i k_B T_i. \qquad (2.48)$$

For the validity of quasi-neutrality the Debye-length λ_D must always be less than the plasma size d. This is assumed to be valid everywhere. Now $\nabla \boldsymbol{v}_{i,e} = 0$ is inserted in eq. (2.45). After that, the quasineutrality condition is inserted, resulting in $\boldsymbol{v}_e = \boldsymbol{v}_i$; both fluids expand with the same velocity, as quasineutrality implies. Even more, this allows to neglect the electron momentum in eq. (2.46) since $\boldsymbol{p}_e \ll \boldsymbol{p}_i$. Hence the electron momentum equation is simply

$$n_e e \boldsymbol{E} = -k_B T_e \nabla n_e. \tag{2.49}$$

The ions are assumed to be initially at rest, i.e., $k_B T_i = 0$. Now the problem is simplified to a one-dimensional case and the electron momentum equation is then inserted into the ion momentum equation. This results in only two equations with two unknowns for the whole fluid:

$$\frac{\partial n}{\partial t} + \frac{\partial (nv)}{\partial z} = 0 \tag{2.50}$$

$$\frac{\partial v}{\partial t} + v_i \frac{\partial v}{\partial z} = -c_s^2 \frac{1}{n} \frac{\partial n}{\partial z}. \tag{2.51}$$

The expansion has a *self-similar solution*[2] [241]:

$$v(z,t) = c_s + \frac{z}{t} \tag{2.52}$$

$$n_e(z,t) = Z n_i(z,t) = n_{e,0} \exp\left(-\frac{z}{c_s t} - 1\right), \tag{2.53}$$

where v denotes the bulk velocity and n_i (n_e) the evolution of the ion (electron) density. The rarefaction wave expands with the sound velocity $c_s^2 = Z k_B T_e / m_i$.
By combining these two equations, replacing the velocity with the kinetic energy $v^2 = 2 E_{\text{kin}}/m$ and taking the derivative with respect to E_{kin}, the ion energy spectrum $\mathrm{d}N/\mathrm{d}E_{\text{kin}}$ from the quasi-neutral solution per unit surface and per unit energy in dependence of the expansion time t is obtained as [67]

$$\frac{\mathrm{d}N}{\mathrm{d}E_{\text{kin}}} = \frac{n_{e,0} c_s t}{\sqrt{2 Z k_B T_{\text{hot}} E_{\text{kin}}}} \exp\left(-\sqrt{\frac{2 E_{\text{kin}}}{Z k_B T_{\text{hot}}}}\right). \tag{2.54}$$

---

[2] A self-similar rarefaction wave keeps the same overall profile at all times, each part of the wave travels at the speed of sound. The solution above is self-similar in the variable $\xi = z/t$.

The ion number N is obtained from the ion density as $N = n_{e,0} c_s t$. Additionally, the electric field in the plasma is obtained from eq. (2.49) as

$$E = \frac{k_B T_e}{e c_s t} = \frac{E_0}{\omega_{pi} t}, \tag{2.55}$$

with $E_0 = (n_{e,0} k_B T_e / \varepsilon_0)^{1/2}$ and $\omega_{pi} = (n_{e,0} Z e^2 / m_i \varepsilon_0)^{1/2}$ denoting the ion plasma frequency (cf. eq. (2.15)). The electric field is uniform in space (i.e. constant) and decays with time as t^{-1}. The temporal scaling of the velocity is obtained by solving the equation of motion $\dot{v} = Zq/m\, E$ with the electric field from above. This yields

$$v(t) = c_s \ln(\omega_{pi} t) + c_s \tag{2.56}$$
$$z(t) = c_s t \left(\ln(\omega_{pi} t) - 1\right) + c_s t. \tag{2.57}$$

Both equations satisfy eq. (2.52). The scaling of the ion density is found as $n(t) = n_0 / \omega_{pi} t$.

However, at $t = 0$, the self-similar solution is not defined and has a singularity. Hence the model of a self-similar expansion is not valid for a description of TNSA at early times and has to be modified. Additionally, in TNSA there are more differences: firstly, the expansion is not driven by an electron distribution being in equilibrium with the ion distribution, but by the relativistic hot electrons that are able to extend in the vacuum region in front of the ions. There quasi-neutrality is strongly violated and a strong electric field will built up, modifying the self-similar expansion solution. Secondly, the initial condition of equal ion and electron densities must be questioned, since the hot electron density with $n_e \approx 10^{20}$ cm^{-3} (eq. (2.22)) is about three orders of magnitude below the solid density of the rear side contamination layers. This argument can only be overcome by the assumption of a *global* quasi-neutrality condition $Z n_i = n_e$. Thirdly, it might not be reasonable to assume a model of an *isothermal* plasma expansion. It can be assumed, however, that the expansion is isothermal for the laser pulse provides "fresh" electrons from the front side, i.e., the assumption is valid as long as the laser pulse duration τ_L. As will be shown below, the main acceleration time period is on the order of the laser pulse duration. This justifies the assumption of an isothermal expansion.

The plasma expansion including charge separation was quantitatively described by Mora et al. [67–71] with high accuracy. The main point of this model is a plasma expansion with charge-separation at the ion front, in contrast to a conventional, self-similar plasma expansion. The plasma consists of electrons and protons, with a step-like initial ion distribution and an electron ensemble being in thermal equilibrium with its potential. The MeV electron temperature results in a charge separation being present for long times. It leads to enhanced ion-acceleration at the front, compared to the case of a normal plasma expansion. This difference is sometimes named the TNSA-effect.

Although being only one-dimensional, the model has been successfully applied to experimental data at more than ten high-intensity short-pulse laser systems worldwide in a recent study [15]. It was separately used to explain measurements taken at the ATLAS-10 at the Max-Planck-Institute in Garching, Germany [59, 242] as well as to explain results obtained at the VULCAN PW [16] (with little modifications). Therefore, it can be seen as a reference model that is currently used worldwide for an explanation of TNSA.

Because of its success in the description of TNSA it will be explained in more detail now. After the laser-acceleration at the foil's front side the electrons arrive at the rear side and escape into vacuum. The atoms are assumed to be instantly field-ionized, leading to $n_i = n_e/Z$. Charge separation occurs and leads to an electric potential Φ, according to Poisson's equation:

$$\varepsilon_0 \frac{\partial^2 \Phi}{\partial z^2} = e\left(n_e(z) - n_i(z)\right). \tag{2.58}$$

The electron density distribution is always assumed to be in local thermal equilibrium with its potential:

$$n_e = n_{e,0} \exp\left(\frac{e\Phi}{k_B T_{\text{hot}}}\right), \tag{2.59}$$

where the electron kinetic energy is replaced by the potential energy $e\Phi$. The initial electron density $n_{e,0}$ is taken from eq. (2.34). The ions are assumed be of initial constant density $n_i = n_{e,0}$, with a sudden drop to zero at the vacuum interface.

The boundary conditions are chosen, so that the solid matter in one half-space ($z \leq 0$) perfectly compensates the electric potential for $z \to -\infty$, whereas for $z \to \infty$ the potential goes to infinity. Its derivative $E = -\partial \Phi/\partial z$ vanishes for $z \to \pm\infty$.

In the vacuum region (initially $z > 0$), the field can be obtained analytically [230]. The resulting potential is

$$\Phi(z) = -\frac{2k_B T_{\text{hot}}}{e} \ln\left(1 + \frac{z}{\sqrt{2\exp(1)}\lambda_{D,0}}\right) - \frac{k_B T_{\text{hot}}}{e} \tag{2.60}$$

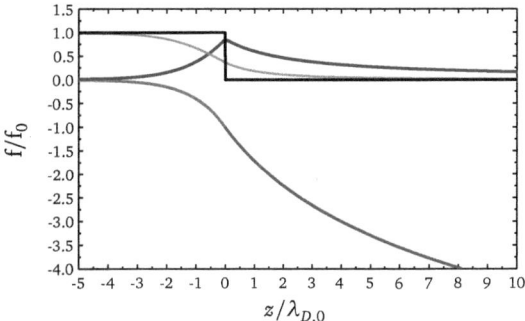

Fig. 2.8: Solution of eq. (2.58). The potential Φ (———) was obtained numerically. The analytical solution eq. (2.60) (———) is in perfect agreement. Both are given in units of $k_B T_{\text{hot}}/e$. The electron density n_e (———), normalized to $n_{e,0}$, follows from eq. (2.59). The normalized ion density n_i (———) is a step-function with $n_i(z < 0)/n_{e,0} = 1$ and zero for $z > 0$. The electric field E (———) is given in units of $k_B T_{\text{hot}}/e\lambda_{D,0}$. The coordinate z is given in units of $\lambda_{D,0}$.

and the corresponding electric field reads

$$E(z) = \frac{2k_B T_{\text{hot}}}{e} \frac{1}{z + \sqrt{2\exp(1)}\lambda_{D,0}}. \tag{2.61}$$

The initial electron Debye length is $\lambda_{D,0}^2 = \varepsilon_0 k_B T_{\text{hot}}/e^2 n_{e,0}$. The full boundary value problem including the ion distribution can only be solved numerically. The result obtained with MATLAB [233] is shown in figure 2.8. The potential Φ (———) is a smooth function and is in perfect agreement with the analytical solution eq. (2.60) (———) in the vacuum region. Both are given in units of $k_B T_{\text{hot}}/e$. The electron density n_e (———), normalized to $n_{e,0}$, follows from eq. (2.59). The normalized ion density n_i (———) is a step-function with $n_i(z < 0)/n_{e,0} = 1$ and zero for $z > 0$. The electric field E (———) has a strong peak at the ion front, with $E_{\text{max}} = \sqrt{2/\exp(1)}\, E_0 = 0.86\, E_0$. The normalization field E_0 is given by $E_0 = k_B T_{\text{hot}}/e\lambda_{D,0}$. The coordinate z was normalized with $\lambda_{D,0}$.

The subsequent plasma expansion into vacuum is described in the framework of a fluid model, governed by the equation of continuity (left) and the momentum balance (right):

$$\frac{\partial n_i}{\partial t} + \frac{\partial (v_i n_i)}{\partial z} = 0 \qquad \frac{\partial v_i}{\partial t} + v_i \frac{\partial v_i}{\partial z} = -\frac{e}{m_p}\frac{\partial \Phi}{\partial z}. \tag{2.62}$$

The full expansion dynamics can only be obtained numerically. Of particular interest is the temporal evolution of the ion distribution and the evolution of the electric field driving the expansion of the bulk. A part of the work described in this thesis was the development of a Lagrangian code in MATLAB that solves eqs. (2.59), (2.60) and (2.62), similar to ref. [67]. The numerical method is similar to the method described in ref. [242], however the code developed here uses MATLAB's built-in bvp4c-function for a numerical solution of the boundary value problem (BVP) in the ion fluid. The initially constant ion distribution is divided into a grid, choosing the left boundary to be $L \gg c_s t$. The boundary value for the potential is $\Phi(-L) = 0$. At the right boundary (initially at $z = 0$) the electric field $-\partial \Phi_{\text{front}}/\partial z = \sqrt{2/e}\, k_B T_{\text{hot}}/e\lambda_{D,\text{front}}$ has to coincide with the analytical solution of eq. (2.61), where the local Debye length has to be determined by the potential at the front:

$$\lambda_{D,\text{front}} = \lambda_{D,0} \exp\left(\frac{e\Phi_{\text{front}}}{k_B T_{\text{hot}}}\right)^{-1/2}. \tag{2.63}$$

Initially, the Debye length at the ion front is obtained by inserting eq. (2.60) in eq. (2.59) to $\lambda_{D,0,\text{front}} = e^{-1}\lambda_{D,0}$.

The code divides the fluid region into a regular grid. Each grid element (cell) has a position z_j and an ion density n_j, as well as a velocity v_j. For each time step Δt, the individual grid elements are moved according to the following scheme [242]:

$$z_{j'} = z_j + v_j \Delta t + \frac{e}{2m_p} E \Delta t^2 \tag{2.64}$$

$$v_{j'} = v_j + \frac{e}{m_p} E \Delta t. \tag{2.65}$$

After that, the density of the cell is changed according to the broadening of the cell due to the movement:

$$n_{j'} = n_j \frac{\Delta x_j}{\Delta x_{j'}}. \tag{2.66}$$

At the front, the individual cells quickly move forward, resulting in a "blow-up" of the cells, that dramatically diminishes the resolution. Thus, after each time-step the calculation grid is mapped onto a new grid ranging from z_{min} to the ion front position z_{front} with an adapted cell spacing. This method is called *rezoning*. The new values of v_j and n_j are obtained by third-order spline interpolation, providing very good accuracy.

Temporal evolution and scaling

A crucial point in the ion expansion is the evolution of the electric field strength E_{front}, the ion velocity v_{front} and the position z_{front} of the ion front. Expressions given by Mora are

$$E_{\text{front}} \simeq \left(\frac{2n_{e,0}k_B T_{\text{hot}}}{e\varepsilon_0} \frac{1}{1+\tau^2}\right)^{1/2}, \tag{2.67}$$

$$v_{\text{front}} \simeq 2c_s \ln\left(\tau + \sqrt{1+\tau^2}\right), \tag{2.68}$$

$$z_{\text{front}} \simeq 2\sqrt{2e}\lambda_{D,0} \left[\tau \ln\left(\tau + \sqrt{1+\tau^2}\right) - \sqrt{1+\tau^2} + 1\right], \tag{2.69}$$

where $e = \exp(1)$ and $\tau = \omega_{pi}t/\sqrt{2e}$. The other variables in these equations are the initial ion density $n_{i,0}$, the ion-acoustic (or sound) velocity $c_s = (Zk_B T_{\text{hot}}/m_i)^{1/2}$, T_{hot} is the hot electron temperature and $\omega_{pi} = (n_{e,0}Ze^2/m_I\varepsilon_0)^{1/2}$ denotes the ion plasma frequency (cf. eq. (2.15)). Due to the charge separation, the ion front expands more than twice as fast as the quasi-neutral solution in eqs. (2.56),(2.57).
From eq. (2.68) the maximum ion energy is given as

$$E_{\max} = 2k_B T_{\text{hot}} \ln^2\left(\tau + \sqrt{1+\tau^2}\right). \tag{2.70}$$

The particle spectrum from Mora's model cannot be given in an analytic form, but it is very close to the spectrum of eq. (2.54), obtained by the self-similar motion of a fully quasi-neutral plasma expanding into vacuum . The phrase *fully quasi-neutral* should point out, that in this solution there is no charge-separation at the ion front, hence there is no peak electric field.

A drawback of the model is the infinitely increasing energy and velocity of the ions with time, which is due to the assumption of an isothermal expansion. Hence a stopping condition has to be defined. An obvious time duration for the stopping condition is the laser pulse duration τ_L. However, as found by Fuchs et al. [15,61], the model can be successfully applied to measured maximum energies and spectra, as well as to PIC simulations, if the calculation is stopped at $\tau_{\text{acc}} = \alpha\left(\tau_L + t_{\min}\right)$. It was found, that for very short pulse durations the acceleration time τ_{acc} tends towards a constant value $t_{\min} = 60\,\text{fs}$, which is the minimum time the energy transfer from the electrons to the ions needs. The variable α takes into account that for lower laser intensities the expansion is slower and the acceleration time has to be increased. It varies linearly from 3 at an intensity of $I_L = 2 \times 10^{18}\,\text{W/cm}^2$ to 1.3 at $I_L = 3 \times 10^{19}\,\text{W/cm}^2$. For higher intensities α stays constant at 1.3. Hence the acceleration time is

$$\tau_{\text{acc}} = \left(-6.07 \times 10^{-20} \times (I_L - 2 \times 10^{18}) + 3\right) \times \left(\tau_L + t_{\min}\right) \quad (2.71)$$

for $I_L \in [2\times 10^{18}, 3\times 10^{19}[\,\text{W/cm}^2$, or $\tau_{\text{acc}} = 1.3\times \left(\tau_L + t_{\min}\right)$ for $I_L \geq 3\times 10^{19}\,\text{W/cm}^2$.

The Lagrangian code was used to simulate proton acceleration with the laser parameters as in the standard case from above, i.e., with the laser intensity $I_L = 10^{19}\,\text{W/cm}^2$, focused to a spot of $r_0 = 10\,\mu\text{m}$ and with a pulse duration of $\tau_L = 600\,\text{fs}$. The acceleration time is then $\tau_{\text{acc}} = 1.67\,\text{ps}$. The target thickness is $d = 20\,\mu\text{m}$, the angular broadening according to eq. (2.30) is $\theta = 42°$ (FWHM) for electrons with the mean energy $k_B T$, determined by eq. (2.28). The resulting initial electron and proton densities at the target's rear side are $n_{(e,i),0} = 1.4 \times 10^{20}\,\text{cm}^{-3}$. With these parameters the electron temperature is $k_B T_e = 0.96\,\text{MeV}$, the initial Debye length is $\lambda_D = 0.61\,\mu\text{m}$ and the sound velocity is $c_s = 9.58 \times 10^6\,\text{m/s}$. The electric field used for normalization is $E_0 = k_B T_{\text{hot}}/e\lambda_D = 1.56 \times 10^{12}\,\text{V/m}$. The ion fluid was initially set up from $z_{\min}/\lambda_D = -80$ to $z = 0$. To test the accuracy, a computation grid of 2000 cells and time steps of $\Delta t = 2.5\,\text{fs}$ were chosen. Later simulations were performed with 500 cells and $\Delta t = 25\,\text{fs}$.

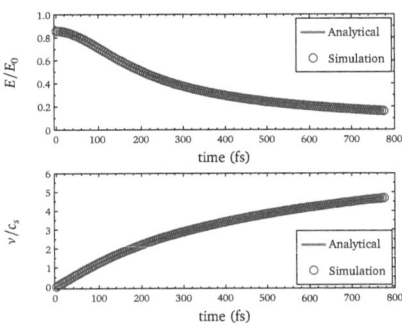

Fig. 2.9: Temporal evolution of the electric field and the ion velocity at the ion front. There is a very good agreement between the simulated values (○) and eqs. (2.67, 2.68) (——).

Figure 2.9 shows the temporal evolution of the electric field and the ion velocity at the ion front, respectively. The electric field was normalized to E_0, the ion velocity is divided by the sound velocity. There is a very good agreement between the simulated values (○) and the expressions by Mora from eqs. (2.67, 2.68) (——). The strongest deviation from the scaling expressions is 1.6 % for the electric field and 0.4 % for the velocity. The electric field evolution, as well as the development of the electron and ion density profiles, are shown in Fig. 2.10. The electric field (——) sharply peaks at the ion front for all times. Initially, the ion density (——) is $n_i = n_0$ for $z \leq 0$ and zero for $z > 0$. The electron density (——) is infinite and decays proportional to z^{-2}. Note the different axes scalings for the electric field and the densities, the latter ones are plotted on a logarithmic scale.

For later times, at $t = (500, 1000, 1500)\,\text{fs}$, the ions are expanded, forming an exponentially decaying profile. A large part of the expanding plasma is quasi-neutral and can be identified by the constant electric field as derived in eq. (2.55). At the ion front, the charge-separation is still present, leading to an enhanced electric field that is a factor of two higher than the electric field in the bulk, in agreement to ref. [67]. This scaling is maintained for the whole expansion. The scaling of the peak electric field value at the ion front with position z, as given by the analytical expressions in eqs. (2.67) and (2.69), is in perfect agreement with the simulation (———).

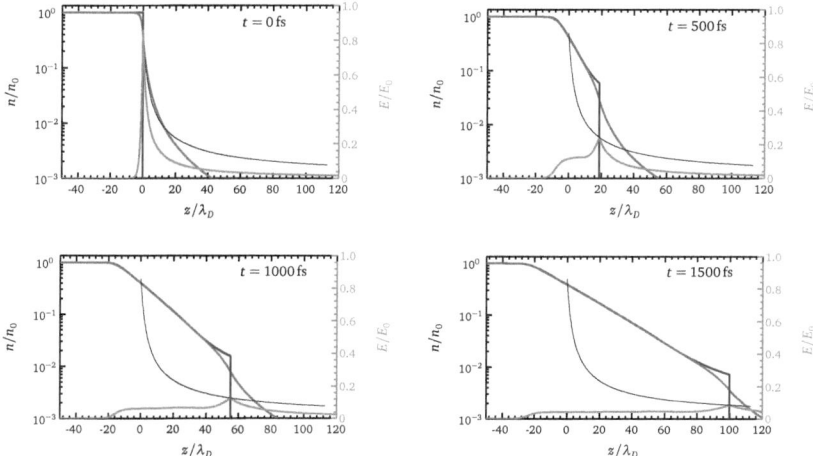

Fig. 2.10: Temporal evolution of the electric field and the ion and electron density, respectively. The electric field (———) sharply peaks at the ion front. The ion density (———) is $n_i = n_0$ for $z \leq 0$ and zero for $z > 0$ for $t = 0$. The electron density (———) decays proportional to z^{-2}. For later times, at $t = (500, 1000, 1500)\,\text{fs}$, the ions are expanded, forming an exponentially decaying profile.

The final proton spectrum is shown in figure 2.11. The numerical solution (———) is close to the analytical one from the quasi-neutral model by eq. (2.54) (———). The analytical spectrum is assumed to reach up to a maximum energy, taken from eq. (2.70). The maximum energy in the simulation is $E_{\text{max,num.}} = 19\,\text{MeV}$, that is in close agreement to the analytical value of $E_{\text{max,analyt.}} = 18.5\,\text{MeV}$. As expected, there is an excellent agreement in the spectra for low energies, since in both cases the expansion is quasi-neutral. For high energies, the numerical spectrum deviates from the self-similar model. The numerical spectrum is lower than the self-similar one even though the ion density of the numerical solution increases close to the ion front, as can be seen in fig. 2.10 in the deviation of the electron and ion densities close

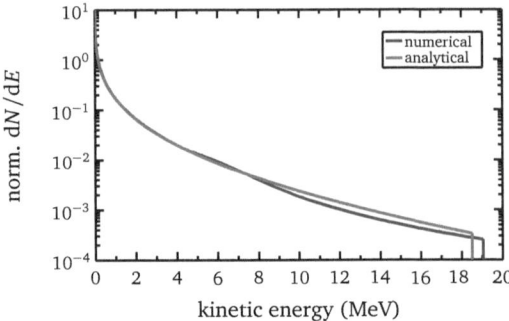

Fig. 2.11: Energy spectrum dN/dE from the simulation (—) compared to the spectrum of a quasi-neutral plasma expansion (—).

to the front. However, the velocity increase at the front in the simulation is much faster than the self-similar solution, due to the enhanced electric field. Thus, the kinetic energy of the fluid elements close to the ion front is higher than the kinetic energy of fluid elements in a self-similar expansion. The spectrum is obtained by taking the derivative of the ion density with respect to the kinetic energy. In turns out, that the kinetic energy increases stronger than the ion density, hence dN/dE is a little less than the self-similar expansion.

In conclusion, the Lagrangian code and the model developed by Mora show, that TNSA-accelerated ions are emitted mainly in form of a quasi-neutral plasma, with a charge-separation at the ion front that leads to an enhanced acceleration compared to the expansion of a completely quasi-neutral plasma. For later times, if $\omega_{pi}t \gg 1$, the analytical expression of the maximum ion energy in eq. (2.70) can be used to accurately determine the cut-off energy of TNSA-accelerated ions. The spectral shape of the ions is close to the spectrum of a quasi-neutral, self-similar expansion.

The equations show, that the maximum energy, as well as the spectral shape, strongly scale with the *hot electron temperature*. The expression for the initial electric field scales as $E \propto \sqrt{k_B T_{\text{hot}} n_e}$, hence a simplistic estimate would assume that both are equally important for the maximum ion energy. In contradiction to that, the investigation has shown that the maximum ion energy only *weakly* depends on the hot electron density and is *directly proportional* to the hot electron temperature. It is worth noting that this finding is in agreement with the (unpublished) results obtained earlier with an electro-static PIC code by Brambrink [243].

The hot electron density - due to the quasi-neutrality boundary condition - determines the number of the generated ions. Both the number of ions as well as the energy are increasing with time, that again shows that not the shortest and most intense laser pulses are favorable for TNSA, but somewhat longer pulses on the order of a picosecond. This requires a high laser energy to keep the intensity sufficiently high.

Nevertheless, the model is still very idealized, since it is one-dimensional and isothermal, with the electrons ranging into infinity and it neglects the laser interaction and electron transport. An approach with electrons in a Maxwellian distribution always leads to the same asymptotic behavior of the ion density [244], hence two-temperature [245] or even tailored [238] electron distributions will lead to different ion distributions. There are many alternative approaches to the one described here, including e.g. an adiabatic expansion [69], multi-temperature effects [69, 245], an approach where an upper integration range is introduced to satisfy the energy conservation for the range of a test electron in the potential [234], the expansion of an initially Gaussian shaped plasma [68] or the expansion of a plasma with an initial density gradient [70]. Most of these approaches assume an underlying fluid model, where particle collisions are neglected and the fluid elements are not allowed to overtake each other. Hence a possible wave-breaking or accumulation of particles is not included in the models but requires a kinetic description, e.g. [71, 246]. Furthermore, the transverse distribution of the accelerated ions cannot be determined from a one-dimensional model and requires further modeling. This will be done in the framework of a two-dimensional particle-in-cell (PIC) simulation with the the Plasma Simulation computer Code PSC, developed by Hartmut Ruhl. A PIC simulation allows a much more sophisticated description, including relativistic laser-plasma interaction, a kinetic treatment of the particles, as well as a fully three-dimensional approach.

2.4.2 Two-dimensional Particle-In-Cell (PIC) simulation

This subsection describes the results of a two-dimensional particle-in-cell (PIC) simulation of a short laser pulse interaction with a dense plasma and the subsequent proton acceleration. Therefore the Plasma Simulation Code PSC, developed by Hartmut Ruhl, has been implemented on a computing cluster. Details about the PSC and its numerical implementation can be found in the appendix, chapter A.

PSC was set up for a 2D simulation in the (y, z)-plane. The simulation box was $(50 \times 50)\,\mu m^2$ in size, consisting of 2500×2500 cells. The spatial resolution is thus 20 nm per cell (50 cells per micron). The boundary conditions were radiating for the fields and periodic in y-direction for the particles. In z-direction the boundary condition was chosen to be reflecting.

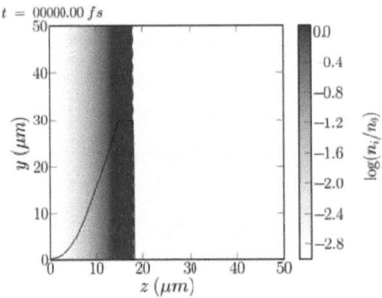

Fig. 2.12: Logarithmic plot of the initial ion density distribution. The front side of the foil decays exponentially, representing the pre-plasma from the pre-pulse. The rear side has a sinusoidal shape. The dashed line represents a line-out along z-direction at $y = 25\,\mu m$.

The $3\,\mu m$ thick target in the simulation consists of gold ions. The rear side is coated with a proton film of thickness $d = 0.5\,\mu m$. Each cell has been filled with 20 quasi-particles per species (electrons and ions). The thickness of the proton coating is slightly higher than the real proton layer thickness, to prevent an unrealistically complete depletion of the rear side during the acceleration. The rear side is modulated by a sine-structure with $3\,\mu m$ period and $0.1\,\mu m$ depth, representing corrugations of the target surface that imprint in the proton beam. This *microfocusing* effect will be further discussed in subsection 3.5.2. At the front side the plasma decays exponentially with a scale length of $1.5\,\mu m$ to mimic the pre-plasma generated by the pre-pulse. The pre-plasma generation takes place on a nanosecond time-scale, therefore it cannot be included in the PIC-simulation. Figure 2.12 shows a plot of the initial ion density. Note that the plot is logarithmically scaled. The line represents a line-out along the z-direction at $y = 25\,\mu m$. The whole target is singly ionized. The initial plasma density is $n_0 = 10^{22}\,cm^{-3}$, the initial electron temperature is $1\,keV/k_B$. The initial ion temperature is $0\,keV/k_B$. Hence the Debye length in the dense part is $2.3\,nm$, that is a factor of ten below the cell size of the simulation box. A larger number of cells in the simulation is far beyond the capabilities of the current cluster. However, the purpose of the simulation is to get further insight into the dynamics of proton acceleration from the rear side. In the previous section discussing the TNSA, the plasma density of the hot electron sheath at the rear side was determined to be more than two orders of magnitude below the initial density (eq. (2.34)) resulting in Debye lengths of 23 nm and more, that can be resolved by the current cell size.

The laser pulse is Gaussian in space and time, with 500 fs FWHM pulse duration (corresponding to a length $dz = 150\,\mu$m) and 4 μm FWHM focal diameter. The laser enters the simulation box at $z = 0$ and $y = 25\,\mu$m. The laser has a wavelength of one micrometer and a peak intensity of $I_0 = 10^{19}\,\text{W/cm}^2$ ($a_0 = 2.7$). This corresponds to a laser energy of 620 mJ. Initially, the peak of the laser pulse is located 225 μm in front of the simulation box. In order to initially have zero laser field in the target, the laser field is set to zero at a distance of $2.5 \times dz$. The simulation runs for about 1 ps, with time steps of 0.03 fs. The laser and target parameters in the simulation have been chosen such that it is essentially as close to reality as possible but still feasible from the perspective of computational cost. The simulation was carried out on 16 processors, the total simulation run needed about four weeks and produced one TB of data.

Results

The simulation can be divided in three stages: i) the laser hits the plasma and generates hot electrons; ii) the electrons penetrate the target and create a strong electric field at the rear side; iii) ion acceleration. These three stages will now be discussed in more detail.

stage 1 - laser-plasma interaction and hot electron generation

As soon as the laser's electric field is strong enough, it creates relativistic electrons with energies above 511 keV. Figure 2.13 shows the laser's electric field (a), the electron density (b), the longitudinal field E_z and the ion density n_i at $t = 320$ fs after the begin of the simulation, respectively. Here and in the following, only normalized data are plotted. The fields are normalized to the laser peak field at $E_0 = 8.7 \times 10^{12}\,\text{V/m}$, the densities are normalized to $n_0 = 10^{22}\,\text{cm}^{-3}$. Note that the fields have been plotted linearly, whereas the densities have been plotted on a logarithmic scale. The dashed lines in the field plots denote the solid density target between $z = 15\,\mu$m and 18 μm. The line-outs are taken at $y = 25\,\mu$m and are averaged over $\pm 2.5\,\mu$m for a better signal-to-noise ratio.

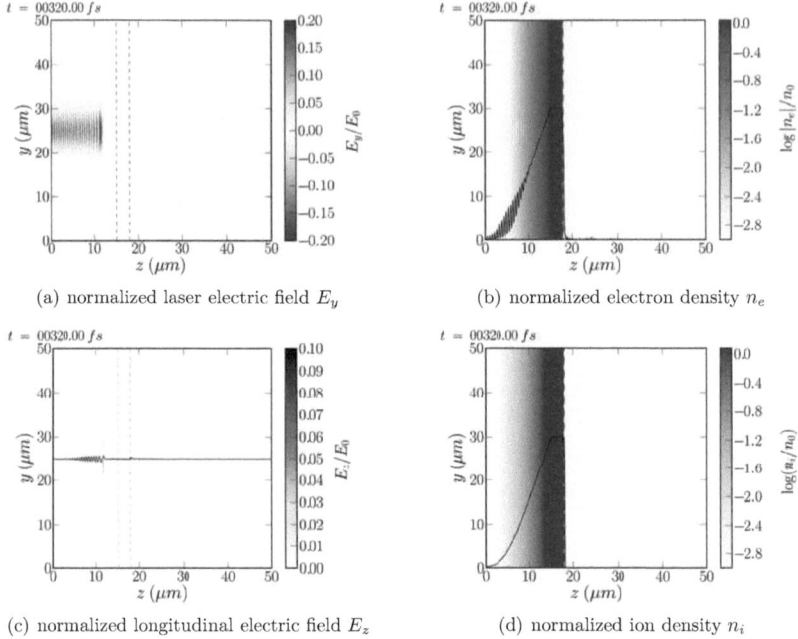

Fig. 2.13: Fields and densities at $t = 320$ fs. The laser is stopped at the critical density (a). The electric field leads to oscillations of the electrons in the pre-plasma (b). This leads to a longitudinal electric field (c), since the ions do not move during this short time (d).

At $t = 320$ fs, the laser pulse has just reached the critical density at $z = 11.5\,\mu$m (a). The oscillating electric field leads to oscillations of the electrons in the pre-plasma (b). This leads to a longitudinal field (c), since the ions (d) do not move in this short time period. A few electrons have started to penetrate the vacuum region at the rear side, since they were set-up with a 1 keV initial temperature. This has lead to the tiny peak in the longitudinal field at the rear side.

Figure 2.14 shows the distribution of accelerated electrons with energies above their rest energy of 511 keV. At $t = 400$ fs (left image) the laser has generated a few electrons that have already started to propagate into the target. Later on, at $t = 440$ fs, a bunched structure of the electrons in the low density pre-plasma appears. The bunches arise at half a wavelength distance, at a frequency of $2\omega_L$, twice every cycle. This is due to relativistic $\boldsymbol{j} \times \boldsymbol{B}$ heating, as described in section 2.2.1. For later times, at $t = 560$ fs, the ponderomotive force has resulted in strongly accelerated electrons in an area with $10\,\mu$m transverse extension. The laser has created a collisionless shock [204], that is clearly visible as the curved region of fast electrons. In addition to that, two jet-like electron beams are injected under 30° angle into the target. This angle has been found in experiments as well, see sec. 2.3.1, however no explanation is given for the reason. The simulation indicates, that the two electron jets may result from resonance absorption in the wings of the laser pulse at the now curved critical density surface. However, a more detailed investigation of the interaction of the laser with the pre-plasma at the front side is beyond the scope of this thesis. The reader is referred to ref. [204], which has investigated the laser-plasma interaction in more detail.

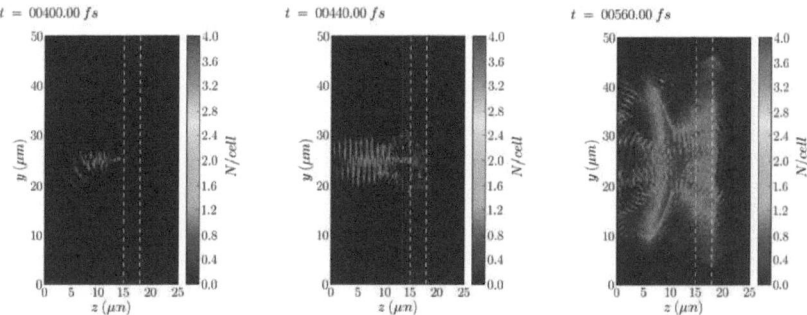

Fig. 2.14: Three snapshots of the distribution of hot electrons with energies above 511 keV. See text for details.

stage 2 - electric field generation

At about 450 fs after the start of the simulation the first electrons have reached the rear side (see fig. 2.14). They enter the vacuum region and generate an electric field. Figure 2.15 shows the electric field (a,b), the electron density (c,d) and the ion density (e,f) for $t = 453$ fs (left column) and $t = 600$ fs, respectively. At 453 fs, the electric field is generated over about $10\,\mu$m in transverse direction, that is close to the size of the laser focal spot. The field immediately spreads along the surface until it reaches the end of the simulation box at $t = 600$ fs. The velocity of this transverse spread is $v = 1.7 \times 10^8$ m/s, half the speed of light. The electron distribution decays exponentially in the vacuum region. This observation justifies the assumption of a thermal, Boltzmann-like electron distribution made in the previous chapters. The ion density still resembles a sharp drop to zero, however they have just started to move in these 150 fs.

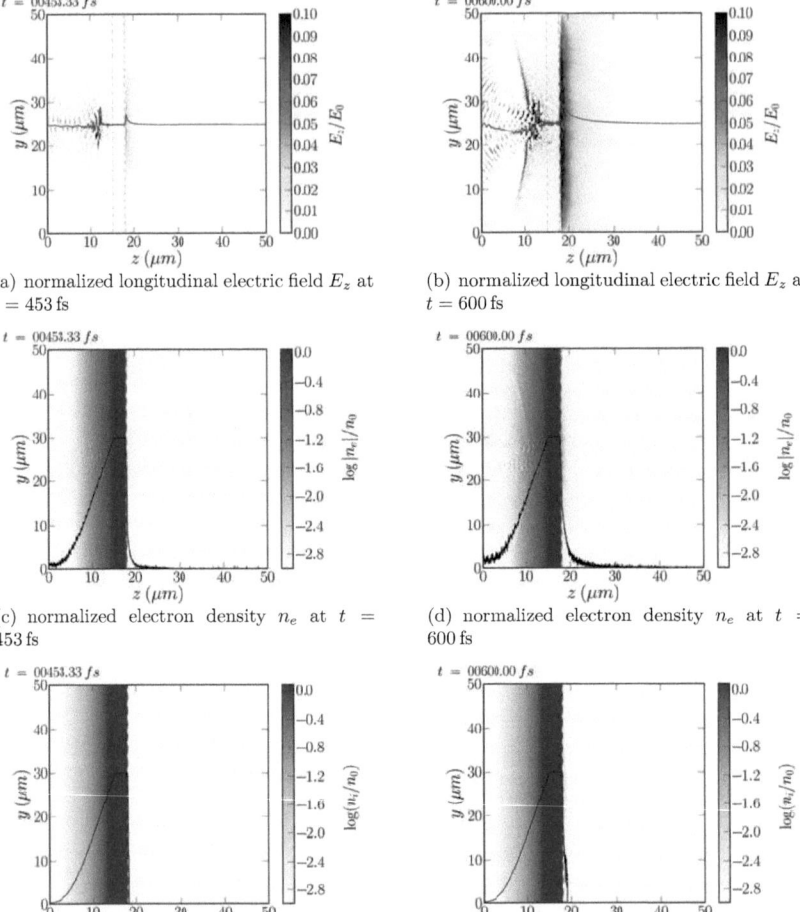

Fig. 2.15: Longitudinal electric field (a,b), electron density (c,d) and ion density (e,f) at $t = 453\,\text{fs}$ and $t = 600\,\text{fs}$, respectively. See text for details.

The peak field at $t = 600\,\text{fs}$ (fig. 2.15b) is about $0.12 \times E_0 = 10^{12}\,\text{V/m}$. The field decays in transverse direction. Figure 2.16 shows a blow-up of the electric field at the rear side of the target. The cyan dots show the positions of half the maximum field amplitude $E_z = 0.06\,E_0$. The transverse shape of the longitudinal electric field can be approximated by a Gaussian (green line) with a FWHM of $18\,\mu\text{m}$ and a $2.2\,\mu\text{m}$ height.

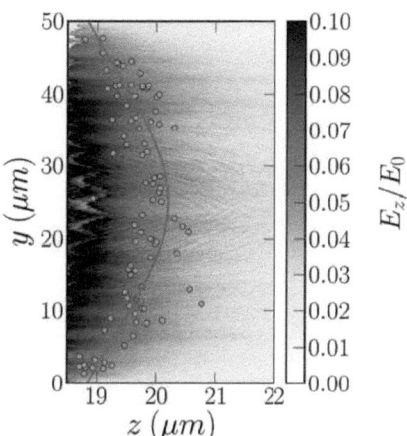

Fig. 2.16: Blow-up of fig. 2.15b.

In summary, the PIC-simulation shows that prior to the ion acceleration, the electrons spread out into the vacuum and over the rear surface. The shape of the hot electron sheath can be well approximated by an exponentially decaying profile in longitudinal direction and a Gaussian or bell-shaped profile in the transverse direction.

stage 3 - ion acceleration

The ion acceleration is driven by the strong electric field created by the electrons. Figure 2.17 shows the temporal evolution of the electric field (left column) and the ion density (right column). (a,b) show the field and density at $t = 600\,\text{fs}$, i.e., at the start of the ion expansion as discussed in the previous subsection. The center plots show the field distribution and ion density at $t = 840\,\text{fs}$, i.e., in the mid of the expansion phase. The electric field is curved, with its maximum extension opposite to the laser impact position. The field not only decays in transverse direction, but also towards the target. The blue line-out was taken at $y = (25\pm2.5)\,\mu\text{m}$. Just as in the case of the fluid expansion in section 2.4.1, the field increases from the vacuum region until it reaches the ion front. There it peaks and starts to decay again. The ion density at the right side is transversely curved as well. In longitudinal direction it decays exponentially, as expected (the line-out is plotted logarithmically). More interesting is the imprinting effect of the sinusoidal rear surface. Each dip has acted like a lens, focusing the protons to a beamlet that expands with the whole ion fluid.

At the end of the simulation, at $t = 1.1\,\text{ps}$, the limits of the simulation box cannot be neglected anymore. In y-direction the boundary leads to a modification of the electric field profile. The enhanced electric field at the ion front is still clearly visible. The field amplitude at the front is $E_z/E_0 = 0.04$, that is a factor of two larger than in the bulk where $E_z/E_0 = 0.02$. This is identical to the plasma expansion with charge-separation, described in section 2.4.1. The plot of the ion density shows the same features as before, however the imprinted grooves are only barely visible due to the limited number of quasi-particles in the simulation. At this point in time, it is futile to proceed further since the poor particle number will result in erroneous data. Another simulation with significantly more quasi-particles, however, is beyond the capabilities of the current computing cluster as it would result in unacceptable computation time (several months).

The plots in fig. 2.17 show a close similarity to the plasma expansion, that will now be more closely investigated. Since the plasma expansion model is one-dimensional, the data from the PIC simulation are taken at $y = (25 \pm 2.5)\,\mu\text{m}$. Only the expansion at the rear side is considered, i.e., the region between $z = 18\,\mu\text{m}$ and $40\,\mu\text{m}$. Figure 2.18 shows the electric field (—), the ion density (—) and the electron density (—), respectively. The three images correspond to the same points in time as the plots in fig. 2.17. The whole time-duration of the expansion is 500 fs. The evolution of the electric field in the PIC simulation is compared to the isothermal fluid expansion from section 2.4.1. Hence the data at $t_{\text{PIC}} = 600\,\text{fs}$ correspond to $t_{\text{fluid}} = 0$, the time $t_{\text{PIC}} = 840\,\text{fs}$ corresponds to $t_{\text{fluid}} \approx 250\,\text{fs}$ and the time $t_{\text{PIC}} = 1100\,\text{fs}$ to $t_{\text{fluid}} \approx 500\,\text{fs}$, respectively. The electric field profiles from the fluid expansion model scale with n_{hot} and $k_B T_{\text{hot}}$, where both quantities are not accurately known from the PIC simulation. Therefore only a similarity can be shown, by fitting the fluid model profiles to the data from the simulation.

The fit at initial time is shown in figure 2.18a. The data from the fluid model are shown as the dashed, black line. In the PIC simulation, the maximum field is $E_z = 10^{12}\,\text{V/m}$. The peak amplitude in the fluid model (see eq. (2.61)) for the standard case is $E = 1.3 \times 10^{12}\,\text{V/m}$, close to the value in the PIC simulation. For a proper fit of the width of the fluid model curve – that is originally given in units of z/λ_D – it had to be multiplied by 0.42, implying only little differences in both cases.

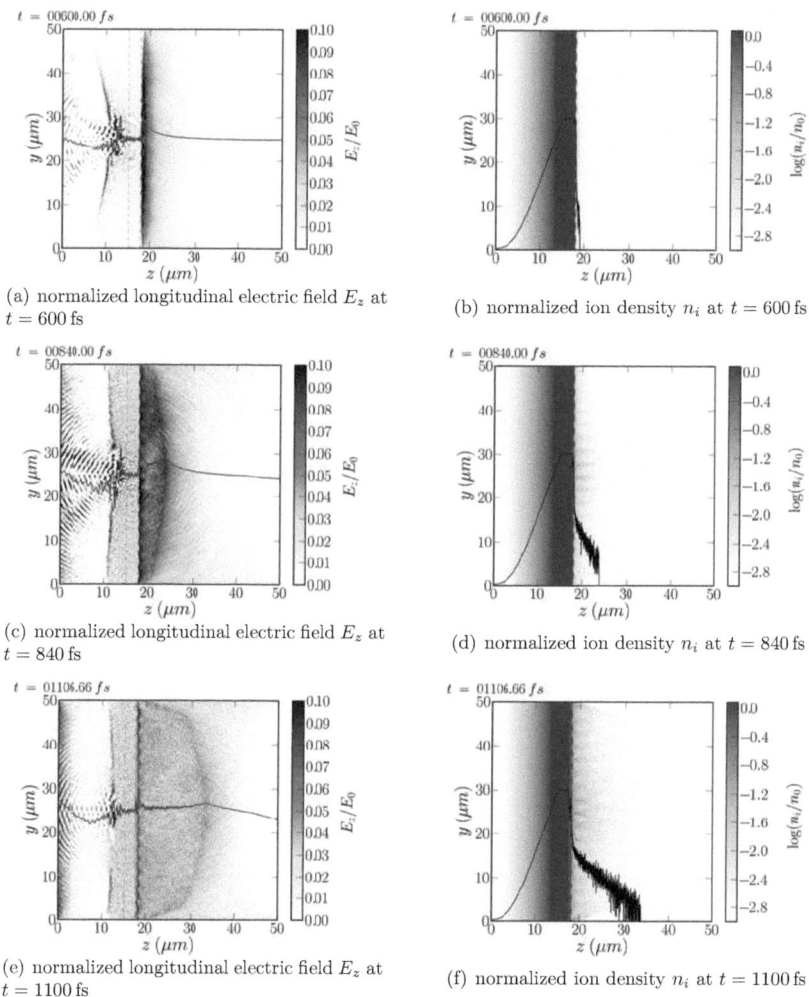

Fig. 2.17: Longitudinal electric field (a,c,d) and ion density (b,e,f) at $t = 600\,\text{fs}$, $840\,\text{fs}$ and $1.1\,\text{ps}$, respectively. See text for details.

The intermediate time step, at $t_\text{PIC} = 840\,\text{fs}$ ($t_\text{fluid} = 250\,\text{fs}$) barely agrees with the fluid model. For the fit the multiplication factor of the width has not been changed. The slope of the electric fields in the vacuum region are close to each other. In the plasma region they strongly differ, with the field in the PIC-simulation being higher

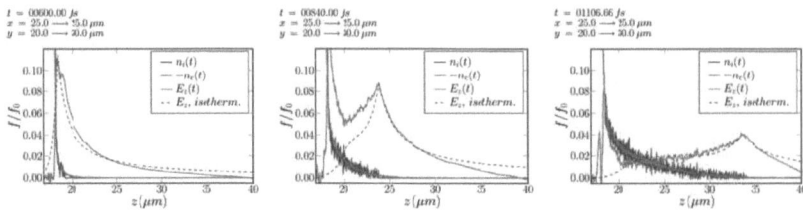

Fig. 2.18: One-dimensional line-outs from the PIC simulation (solid lines) versus electric field from the fluid model (dashed line). See text for details.

than in the isothermal fluid model. Close to the initial surface the electric field in the PIC simulation is still very high, in contrast to the field in the isothermal fluid model. This discrepancy is due to electron kinetic effects during the expansion [71], that cannot be covered by the fluid model.

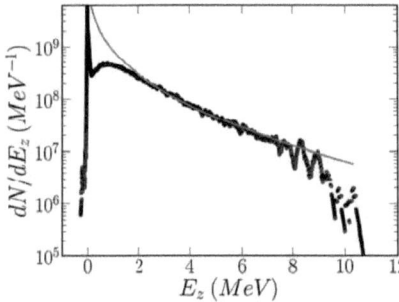

Fig. 2.19: Proton spectrum obtained in the PIC simulation (dots) compared to the spectrum of a quasi-neutral, isothermal plasma expansion (—).

At the end of the simulation run, the fields in the PIC-simulation and the fluid model show much better agreement. Again, the data from the fluid model have been shrinked by a factor of 0.42 in z-direction as above. Except for the region close to the remaining target, the electric field in the PIC simulation can be well approximated by the isothermal plasma expansion model. The discrepancy in the vacuum region is due to returning electrons that have been reflected at the boundary in the PIC simulation at $z = 50\,\mu$m, leading to an artificially strong decay of the field.

The comparison between field dynamics of the two-dimensional PIC simulation and the isothermal fluid model has shown, that the general features of TNSA can be covered by the fluid expansion model. However, for early times below 250 fs the fields strongly deviate from each other, which is most likely due to kinetic effects that cannot be covered by the fluid model [71].

The last point in this discussion is the comparison of the proton spectrum from the PIC simulation to the fluid model. Fig. 2.19 shows the spectrum obtained in the

PIC simulation (dots). The red line is a best fit of eq. (2.54), where $N = n_{e,0}c_s t$ and $Z = 1$. The fit parameters are $N = 3.82 \times 10^9$ and $k_B T = 0.68\,\text{MeV}$, respectively. The quasi-neutral plasma expansion spectrum roughly fits the intermediate part of the spectrum between 2 and 8 MeV. For lower energies, the finite width of the proton layer leads to less protons compared to the isothermal plasma expansion model [69]. At energies above 8 MeV the spectrum from the simulation shows strong modulations, due to the insufficient quasi-particle number. However, the number clearly decays stronger than expected from the plasma expansion model. The discrepancy at higher energies was also found in the experiments and will be further discussed in section 4.1.2.

Chapter 3

Experimental method

3.1 General set-up and laser systems

The experiments have been performed at high-energy, high-intensity laser systems, operating with the chirped-pulse-amplification (CPA) technique [14]. A low-energy, short pulse ($\approx 100\,\text{fs}$) from a fs-laser oscillator is temporally stretched by orders of magnitude up to nanosecond pulse duration. The stretched pulse is then amplified up to ten's of Joules in several glass-based amplifier stages and is then re-compressed below 1 ps pulse duration in a vacuum vessel. Vacuum is necessary to avoid non-linear interactions with the air that would compromise the pulse propagation. The pulse is then guided to the experimental vacuum chamber.

The typical experimental set-up will be described with the configuration at the PHELIX at GSI as an example. Details of the PHELIX laser system itself can be found in refs. [243, 247]. Figure 3.1 shows a photograph of the target chamber interior at PHELIX.

The CPA-laser pulse (—) enters the chamber from the right side, coming from the vacuum grating compressor. Three mirrors guide the laser pulse to the off-axis parabolic mirror (OAP), that focuses the pulse onto the target in the target chamber center. Figure 3.2

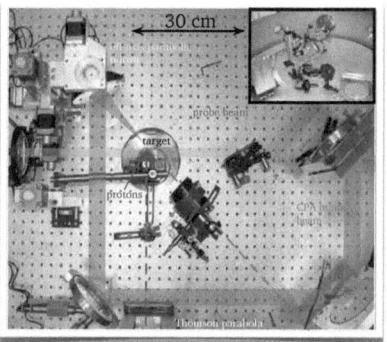

Fig. 3.1: Experimental set-up, the figure is explained in the text.

shows the focal spot of PHELIX. The OAP focused the beam to a nearly diffraction limited spot of 7 μm FWHM. At low intensity (in the dark blue region) the first Airy ring is visible.

A low-energy fraction of the main pulse is used as a probe beam (   ), propagating along the target surface. This beam is used before the experiment to accurately align the target in the focal plane. During the shot, it can be used as an interferometric probe to determine the pre-plasma electron density [180]. Without target, the focal spot of the OAP is imaged (- - -) by an objective to a charge-coupled device (CCD) camera mounted outside the target chamber.

Protons are being accelerated from the target rear side, depicted by the blue triangle. In a distance of a few cm (typically 2 cm up to 5 cm) a film detector is placed. In a few experiments, and in this example as well, a hole or a slit has been machined in the center of the RCF, to allow the ions to propagate up to another detector. In figure 3.1 a Thomson parabola ion spectrometer has been placed in the chamber for a higher spectral resolution and ion species discrimination. A Thomson parabola is described further in refs. [8, 248]. A pinhole has been placed in the beam path (- - -) to collimate the beam for the spectrometer. In most cases, the spectrometer is housed in its own vacuum chamber outside the main chamber to protect it from target debris.

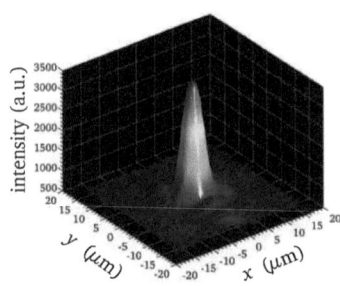

Fig. 3.2: Focal spot of PHELIX.

The experiments have been performed at four different high-intensity CPA-laser systems. Their properties are summarized in table 3.1. All laser systems have Nd-doped glass for the main amplifiers. This allows for large diameter beams to keep the beam intensity below the threshold for non-linear effects, but it diminishes the minimum pulse duration to several hundred fs. All laser-systems have a Ti:Sa oscillator, forced to emit a pulse at a wavelength of 1054 nm for an amplification in the glass amplifiers. The lasers are being operated in a single-shot mode. Due to the long cooling time of the flashlamp-pumped glass amplifiers on the order of 20 min. up to 2 h, the repetition rate was on the order of a few shots a day, resulting in sub-optimum statistics.

Whereas the experiments at TRIDENT and at LULI have been performed with the full system, the experiments at PHELIX have been done with the pre-amplifier section only. The experiments at Z-Petawatt were carried out with a sub-aperture beam at the 100 TW target chamber section. The laser systems had an energy of a few tens of Joules, spot diameters about 10 μm and pulse duration below 1 ps. Hence the peak laser power has been above 10 TW, with peak intensities above 10^{19} W/cm^2, assuming that *all* energy is in the focal spot. The normalized laser amplitude was above one, thus the electrons have been accelerated to relativistic energies during the laser-plasma interaction.

laser system location	PHELIX GSI	LULI-100 TW LULI (F)	TRIDENT LANL (USA)	Z-Petawatt SNL (USA)
energy on target (J)	3	15	25	40
pulse duration (fs)	900	350	600	600
FWHM diameter (μm)	12	6	12	5
peak power (TW)	3	43	42	50
peak intensity (W/cm^2)	3×10^{18}	1.5×10^{20}	3.6×10^{19}	3×10^{20}
norm. amplitude a_L	1.5	10	5	13.5
pulse contrast	10^{-5}	10^{-6}	10^{-6}	10^{-7}
pre-pulse	1 ns	500 ps	2 ns	> 100 ps

Tab. 3.1: Overview of glass laser-systems used for the experiments.

3.2 Ion beam detectors

The detectors used in high-power, high-intensity laser-matter interaction experiments are somehow specialized detectors that must be able to survive the huge electromagnetic noise during the laser-plasma interaction. The intense, short-pulsed ion beams generated in the experiments require novel detection techniques that differ from ion beam detectors used in conventional ion accelerators, as e.g. ionization chambers, Faraday cups, diamond detectors, scintillators, thermoluminescence dosimeters (TLD) or diodes [249, 250]. Ionisation chambers, diamond detectors, Farady cups and diodes have poor spatial resolution and suffer from the strong electromagnetic noise generated in laser-plasma interaction. TLDs can only provide an absolute energy value with very weak spatial resolution due to the limited detector size. Scintillators can be used to determine the beam shape and position, respectively. However, no information can be obtained about the energy distribution. Since the accelerated ions in TNSA have relatively high energies, nuclear activation via (p, n)-reactions can be used to determine the particle number and spectrum, respectively [21-23]. On top of the radiation safety issues this technique has the drawback that it only provides one-dimensional information. Conventional film detectors on the other hand measure the dose distribution in two dimensions, but they are sensitive to light and must be developed after irradiation, that can lead to inaccuracies in the dose determination and consequently in the particle spectrum and number. Nevertheless, film detectors have high spatial resolution and are insensitive to electromagnetic noise. They are available in arbitrary sizes and they are thin, which allows to stack several films. The penetration depth of the ions then allows to determine the ion energy. The film optical density after irradiation is proportional to the dose, that in turn is proportional to the number of ions.

For the detection of laser-accelerated ions the self-developing GafChromic® RadioChromic Film (RCF) [251] in a stack configuration have been proven to work very well. The films have been calibrated already in an earlier work [243], however a different scanner had been used. Additionally, the manufacturer has changed the film composition in the meantime, requiring a new calibration for protons.

The detector itself, as well as the calibration for protons and the measurement technique will be described in the next sections. A more detailed summary on the calibration can be found in refs. [252, 253].

3.3 RadioChromic Film – RCF

GafChromic® RadioChromic Film (RCF) [251] of the types HD-810, HS, and MD-55 are widely used in radiographic imaging, for dosimetry in radiation therapy and in industrial quality control [255, 256]. RCF are radiation-dose sensitive films, consisting of an active monomer that upon exposure to ionizing radiation polymerizes to form a darker dye. Hence no film development is required. The color of the film changes from transparent to different shades of blue, depending on the amount of radiation (dose) that was absorbed in the film.

The diacetylene monomers in the active layer of the film undergo a radiation-induced, auto-catalytic, slow 1,4-trans-polymerization. This results in a change in the absorption spectrum, leading to the change in color [255]. The films are used for dose measurements of energetic photons (γ- or x-rays) or of corpuscular radiation like neutrons, electrons or ions up to doses on the order of 10^5 Gy [257]. The sensitivity range given by the manufacturer is (2-100) Gy for MD-55 and (10-400) Gy for the HD-810. However, different ranges can be found in the literature [257–260], since the limit depends on the type of scanner used for electronic data acquisition. The

MD-55 is used for moderate doses, whereas the HD-810 is suitable for high doses, as the abbreviations suggest. The dose limit will be further described in section 3.4.2.

An enormous advantage of RCF is its high spatial resolution of the absorbed dose up to the micrometer range, due to its small, approximately $(2 \times 2)\,\mu m^2$ grain size of the radiation-sensitive components [254]. A scanning electron microscope image of the radiation-sensitive layer of MD-55 is shown in figure 3.3. The image was taken from ref. [254]. The spatial resolution according to McLaughlin et al. [259] is about $1\,\mu m$.

A disadvantage is its sensitivity against ultra-violet light below 300 nm as e.g. sunlight, that could colorize the film. Mechanical stress like dust, scratches

Fig. 3.3: Scanning electron microscope image of the sensitive layer in *GafChromic®* MD-55 [254].

or finger prints on the film surface can directly affect the coloration process [249]. Material damage can be recognized in a color-change from transparent to milky-white [258]. The film maintains its color up to temperatures of 60° C, above that the film suddenly changes its color from blue to red, making quantitative data analysis impossible. Therefore it is important to store irradiated as well as non-irradiated films in a light-proof container and at room temperature [259].

3.3.1 Film composition

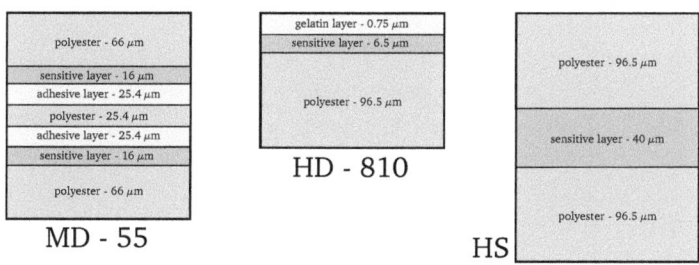

Fig. 3.4: Layer configuration of GafChromic® film types MD-55, HD-810 and HS.

All three types of GafChromic® RCF are composed of different layers. The supporting polyester layer is coated with the radiation sensitive layer that consists of the organic, micro-crystalline monomers in a gelatin substance [255, 258, 260]. The most often used type MD-55 has two active layers, that are glued together by three layers of an adhesive dye and polyester. Figure 3.4 shows a schematic configuration of the different layers and their thicknesses. The chemical composition and densities of the different layers, as provided by the manufacturer [251], are given in table 3.2. The polyester layers are provided for mechanical stability. The least sensitive film, HD-810, has a very thin active layer. On top of the active layer is a thin gelatin layer for protection. Since the end of 2007, the manufacturer delivers new types of HD-810 without the gelatin layer. This is advantageous for the detection of heavy ions, that cannot penetrate the RCF very deeply due to their high stopping power. The most sensitive film HS has a very thick active layer enclosed in polyester. Due to its thick layers, this film is fragile and must be handled carefully. Otherwise the layers will split, leading to a white discoloration.

	density (g/cm^3)	C (%)	H (%)	O (%)	N (%)
polyester	1.35	45.44	36.36	18.20	0.00
sensitive layer	1.08	29.14	56.80	7.12	6.94
adhesive	1.20	33.33	57.14	9.53	0.00
gelatine coating	1.20	22.61	53.52	11.12	12.75

Tab. 3.2: Chemical composition and density of GafChromic® radiochromic films.

3.3.2 Radio-chemical reaction

Upon irradiation with ionizing radiation, the films are mostly transparent with a light blue color. The impact of radiation leads to a polymerization that changs the absorption spectrum. The polymer develops absorption bands at 618 nm and 676 nm, as shown in figure 3.5 [254]. The figure shows the change in optical density[1] OD versus the wavelength. The absorption is proportional to the amount of the generated polymers, that in turn is proportional to the amount of radiation (the energy dose) absorbed in the active layer. The result is visible directly after the experiment and it has the advantage that no further chemical, thermal or optical processing is necessary, minimizing errors in the dose determination.

---

[1] The unitless optical density is defined as the transmittance of an optical element, measured as the logarithm of the transmitted intensity I over the incident intensity I_0 at wavelength λ: $OD_\lambda = \log_{10}\left(\frac{I}{I_0}\right)$.

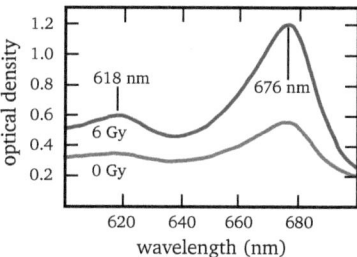

Fig. 3.5: Absorption spectrum of an RCF before (—) and after (—) irradiation with a dose of 6 Gy. Plot according to [254].

About 90 % of the coloration takes place in the first milli-seconds after irradiation. It takes between minutes and hours until the chemical coloration is finished. In the first 24 hours after irradiation the optical density increases up to 16 % its initial value, followed by a two week long increase by 4 % [257–260]. After that, the color stays constant for years.

3.4 RCF calibration for protons

The films have to be digitized for quantitative RCF data analysis. This can be done by transmission-densitometers, spectro-photometers or film-scanners. The scanner converts the radiochromic film to a pixel sequence, with each pixel value representing an absorption value or optical density. The response of the RCF is strongest for scanners with a red light source, i.e., when the light source emits at a wavelength where the films have their maxima of absorption at 618 nm and 676 nm. The drawback is that the film therefore has a low saturation threshold at this wavelength. Hence the film can be used for low optical densities only, or for the detection of low doses. This drawback could be overcome by using different colors for scanning [261], on the cost of a more complex data analysis. For high doses, and therefore high optical densities, the intensity of the light source has to be increased [258]. Another prerequisite is a high dynamic range in the color depth, hence the scanning device should be able to resolve the colors with 16 Bit (65535 colors).

A two-dimensional data acquisition can be performed with two different techniques [262]. The first method, used e.g. in micro-densitometers, is to use a light source and a microscope objective to scan the film spot-by-spot [263]. This method prevents the influence of stray light and can be done with a calibrated light source, e.g. a HeNe-Laser. However, the scan process is very slow and the advantage of a fast data acquisition by the self-developing films is lost. Furthermore, micro-densitometers are not very portable and are very expensive. The second method is to image the whole film with a camera system and a large-area light source, as it is used in conventional flat bed document scanners. This technique is very fast, however the spatial resolution as well as the dynamic range is not as high as in the first method. During a usual experimental campaign hundreds of RCFs are being irradiated. This large amount of data, the flexibility and the possibility that it is widely available has favored the use of a conventional flat bed scanner, even though not the maximum resolution is obtained.

A calibration of the RCF data acquisition system not only requires the films to be calibrated for their response to protons, but the scanner has to be calibrated as well. This includes to test the flat bed scanner for illumination homogeneity, for its temperature stability, reproducibility and its linearity, just to name a few.

3.4.1 Scanner calibration

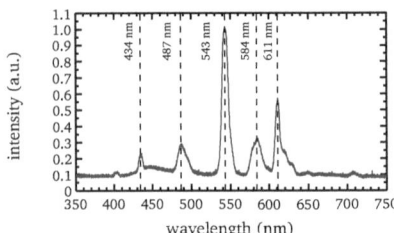

Fig. 3.6: Fluorescent lamp spectrum of the Microtek ArtixScan 1800f.

The manufacturer of RCF recommends the industrial scanner Epson 1600, that operates at a wavelength of 610 nm. Laser-accelerated ions are being generated in a huge number, resulting in large optical densities above $OD = 2$ in the detector. This could be close to the saturation threshold for the red wavelength. Therefore, our group has decided to use a conventional film scanner, the *ArtixScan 1800f* from Microtek (http://microtek.com/). The scanner can both scan in reflection as well as in transmission mode. The scanner is operated in a grey color scan mode, since at the grey color level it is less sensitive and higher doses can be resolved [261]. Its maximum optical density is $OD_{max} = 3.1$, that is enough to resolve even the first RCF layers with the highest optical densities from a usual experimental measurement. The scanner is operated with the software SilverFast Ai V.6 from LaserSoft Imaging (http://www.silverfast.com). Since this scanner is relatively cheap (market price about 1000 €) and since our group has accurately calibrated the RCFs for protons during this thesis, the scanner is now in use at different laser laboratories worldwide (e.g. LULI, LANL, SNL, Universität Dortmund).

The scanner operates with a white-light lamp. The spectrum (behind the glass window) was measured with an OceanOptics HR 4000 spectrometer and is shown in figure 3.6. It consists of several discrete radiation bands that are typical for commercial mercury fluorescent lamps. The most prominent peaks are from terbium- and europium-doped phosphors [264].

A calibrated system requires to be operated with always the same settings and conditions. It was found that the scanner lamp must heat up for about 15 minutes for stable light conditions. The optical density also depends on the temperature, therefore the readings should be performed at room temperature only. The software allows for a large variety of settings; the scanner is operated for the calibration run and for the quantitative RCF analysis with the following software settings:

scan mode	normal	raster	667
original	transmission	resolution	1000 dpi
pos./neg.	positive	filter (blind color)	white
scan type	16 Bit grey scale	lamp brightness	0
filter	none	HiRePP	☑
image type	standard	Q-factor	1.5

The scanner was further calibrated for its optical density with a calibrated and certified grey-color transmission step wedge (part #T4110cc) from *Stouffer Industries, Inc.* [265]. The step wedge has 41 steps in optical density from $OD = 0.06$ to $OD = 4.03$ in step intervals of ΔOD about 0.1. After scanning, each step area was averaged and the resulting pixel value was then plotted versus its corresponding optical density. The plot is shown in figure 3.7 with the open circles (○). The pixel values are converted to optical density by the following relation (—):

$$OD = P_1 + \frac{P_2 - P_1}{1 + \left(\frac{\text{pixel value}}{P_3}\right)^{P_4}}, \qquad (3.1)$$

with the parameters $P_1 = 2731.45$, $P_2 = -0.716$, $P_3 = 0.002$ and $P_4 = -0.468$, respectively. The maximum resolvable optical density is close to $OD_\mathrm{max} = 3$, however due to the strong increase in slope for optical densities above 2.5, the maximum optical density for practical reasons is $OD = 2.5$. This is still sufficient for most data obtained in laser ion acceleration experiments. The upper limit of RCF is about $OD = 5$ as obtained by Bishop et al. [257].

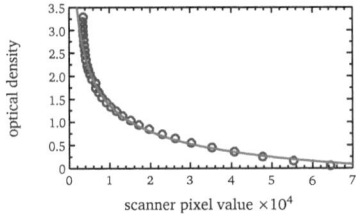

Fig. 3.7: Optical density versus scanner pixel values. The measured values (◦) are fit by eq. (3.1) (—).

3.4.2 RCF sensitive layer calibration

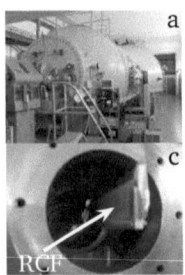

Fig. 3.8: Experimental set-up for the calibration. a) shows the Tandem linear accelerator at MPI Heidelberg, b) shows the beam line and vacuum chamber where the measurement was performed and c) shows an RCF mounted in a frame before irradiation.

The calibration for protons was performed at the Max-Planck-Institut für Kernphysik at Heidelberg, Germany. The RCF were irradiated with 8 MeV energy protons at a current of 50 pA from the Tandem linear accelerator, shown in figure 3.8a. At the interaction point with the RCF, shown in figure 3.8b and c, the proton beam was strongly de-focused. A 1 cm diameter aperture was placed 10 cm before the RCF, cutting most of the beam. The resulting beam profile was very homogeneous with a top-hat profile with less than 5 % deviation. The 8 MeV energy of the protons was chosen since at this energy, the ions can penetrate the film and are not stopped. The proton current was measured with a 1 cm diameter Faraday cup behind the films. The beam was switched on and off by opening and closing a valve.

The energy deposited in the RCF is directly proportional to the amount of charge passing through the film. This could be measured by multiplying the beam current by the irradiation time. However, this simple current measurement has the drawback that current fluctuations in the accelerator (that have been on the order of 5 %) are averaged and, even worse, the switching time (about 1 s) of the valve leads to large errors in the irradiation time, especially for short times. Therefore, the current signal from the Faraday cup was fed into a current digitizer. This device transforms the current amplitude to a frequency signal that is directly proportional to the current. The frequency signal was then plugged to a discriminator and then into a digital counter. This set-up was then calibrated in long-time measurements (without the RCF in place) with a beam current of 500 pA and irradiation times from 20 s to 300 s, leading to linear relation between the charge passing to the Faraday cup and the number of counts in the counter. The data was fit with a linear regression. The χ^2-error of the fit is below 1 %, hence even for very short times the charge could be accurately determined.

Fig. 3.9: MD-55 after irradiation with 8 MeV protons at 50 pA current. From left to right the irradiation times were 2 s, 5 s, 75 s, 130 s, 250 s and 500 s, respectively.

For the RCF calibration measurements the irradiation time was changed from film to film in the interval from 1 s up to 500 s, thereby changing the dose in the active layer. The calibration was performed for all three RCF types HD-810, MD-55 and HS [253]. Figure 3.9 shows an MD-55 after the exposure to the proton beam. The irradiation times for this case were 2 s, 5 s, 75 s, 130 s, 250 s and 500 s, respectively. The proton beam spot is very smooth and round, with less than 5 % fluctuation in dose. The spots from long irradiation times show some background signal from scattered protons at the aperture. However, as the RCF response to protons is strongly non-linear as it will be shown below, this background signal can be neglected.

The placement of the RCF in the beam path leads to scattering in the films as well, diminishing the signal in the Faraday cup behind the film up to a factor of 1/3. Therefore a separate transmission measurement was performed before each film scan by comparing the detector current with and without the RCF in place, to correct for this particle loss behind the RCF.

An ion propagating in matter looses its energy by interacting with the atoms. The differential energy loss or *stopping power*, i.e., the infinitesimal energy fraction dE lost by the particle on its infinitesimal track fraction dx, is described by the Bohr-Bethe-Bloch equation, with some correction factors depending on the initial energy and projectile type. A summary is given in ref. [266]. A unique property of the ion energy loss characteristics is the increase of dE/dx towards the end of the range, with a global maximum just at the end where the particle is stopped. This so-called *Bragg-Peak* is a result of the projectile velocity and target electron velocity in the atom shell approaching the same value, enhancing the energy transfer from projectile to target.

For nowadays TNSA-accelerated proton energies the semi-empirical Stopping and Range of Ions in Matter (SRIM) code package [267] can be used to calculate the differential energy loss dE/dx. SRIM outputs data tables with dE/dx in dependence on the initial energy. Figure 3.10 shows a SRIM calculation by F. Nürnberg for protons penetrating a HD-810 [9]. Four sample initial energies of $E = 0.025\,\mathrm{MeV}$, $E = 0.3\,\mathrm{MeV}$, $E = 1\,\mathrm{MeV}$ and $E = 4\,\mathrm{MeV}$ have been chosen. Plotted is the differential energy loss dE/dx versus the penetration depth. In the background, the different layers of HD-810 have been color-coded in analogy to figure 3.4. The energy loss changes each time the proton enters a different layer. At the end of the range, the energy loss peaks at the Bragg-peak.

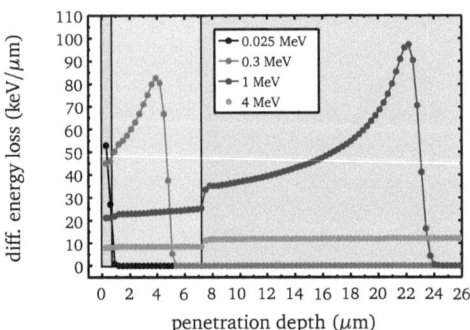

Fig. 3.10: The energy loss of protons in HD-810 has been calculated by F. Nürnberg for four different initial energies (0.025 MeV, 0.3 MeV, 1 MeV and 4 MeV, respectively) with SRIM2006 [267]. The plot shows the differential energy loss dE/dx versus the penetration depth. In the background, the different layers of HD-810 have been color-coded in analogy to figure 3.4. The energy loss changes each time the proton enters a different layer. At the end of the range the energy loss has a maximum (Bragg-peak).

The total energy deposition of each proton in the active layer has been calculated with the SRIM2006 data tables and a ray-tracing algorithm written in MATLAB. The ray-tracing code takes the initial energy ($E_0 = 8\,\text{MeV}$) and calculates the remaining energy after a certain track distance dx by table-lookup and linear interpolation. The RCF consists of different layers with different materials and thicknesses, according to figure 3.4. For the calculation of the integrated energy deposition in the active layer, each layer of a single RCF is sliced into pieces of length x. The initial energy is taken and the energy loss is calculated after passing the slice of length x. After that, the proton has the energy $E_0 - \frac{\Delta E}{\Delta x} x$ which is used as input energy for the next slice. When the current slice is part of the active layer, the deposited energy is summed.

The temporally integrated, deposited energy per mm^2 in the active layer for the calibration measurement is calculated as

$$E_{dep} = \frac{Q\, E_{\text{single}}}{A\, e}, \tag{3.2}$$

with the temporally integrated beam charge Q, the beam area A, the proton charge e and the integrated differential energy loss E_{single} for a single proton calculated by the ray-tracing code. The single particle energy deposition is $E_{\text{single}} = 39.7\,\text{keV}$ for HD-810, $E_{\text{single}} = 211.26\,\text{keV}$ for MD-55 and $E_{\text{single}} = 290\,\text{keV}$ for the HS, respectively. In most publications not the deposited energy, but the energy dose is given. The energy dose is the energy deposited in the active layer per unit mass. Hence it can be calculated as

$$D = \frac{E_{dep}}{d\,\rho} = \frac{Q\, E_{\text{single}}}{e\, A\, d\, \rho}, \tag{3.3}$$

with the active layer thickness d (see figure 3.4) and density ρ from table 3.2

Analogous to the scanner calibration with the step wedge, the irradiated RCFs were first scanned for their optical density with the calibrated ArtixScan 1800f. A MATLAB routine with graphical user interface was written to subtract the background, to cut out the circular irradiation spots and to average over the spot size. The background is negligible for the HD-810. For MD-55 it is about $OD = 0.2$, similar to the values found in ref. [254]. The HS has a background optical density of $OD = 0.2$ as well. The measured optical density is then plotted versus the deposited energy or versus the dose.

Figure 3.11(left) shows the energy deposited in the active layer versus the optical density for all three film types HD-810 (○), MD-55 (□) and HS (▽). The bi-

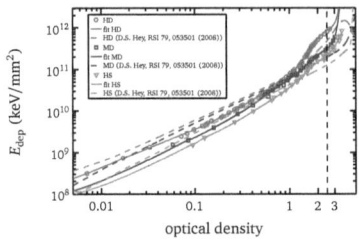 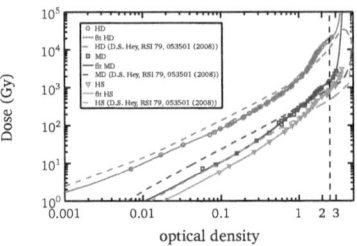

Fig. 3.11: RCF calibration curves. Shown is the energy deposition E_{dep} in keV/mm² (left) and the energy dose in Gy (right) versus the film optical density. The measured data are plotted as open circles. The data have been fit with a ninth-order polynomial from eq. (3.4) (lines). The data for the MD-55 and HS have been extrapolated by eqs. (3.5) and (3.6) for optical densities below $OD = 0.1$ (lines). The data for the HD-810 are shown in red color, the data for the MD-55 are plotted in blue and the data for the HS are plotted in green. The black dashed line denotes the scanner's practical optical density limit of $OD = 2.5$. The dashed lines show the results obtained in ref. [263] with a different scanner and a blue filter for comparison.

logarithmic plot shows a strongly non-linear slope that saturates at optical densities around $OD = 3$ due to the scanner saturation. The black dashed line denotes the practical optical density limit of $OD = 2.5$. The error bars are not plotted, since they are on the order of the symbol size.

The curves are similar for each RCF type and differ only in the vertical position. Ideally, the curves are identical, since the films all have the same sensitive layer. However, due to the different sensitive layer thicknesses, with the HD having the thinnest sensitive layer and the HS having the thickest one, the measurement shows that a volume effect is present. This means, that for obtaining the same optical density in an RCF the necessary energy deposition depends on the RCF type. This issue is even more visible in the dose plot on the right side, since for the dose determination the irradiated volume is explicitly taken into account. Comparing the energy deposition necessary to obtain a certain optical density, the calibration has shown that HD-810 is the least sensitive film and HS is the most sensitve film as expected. However, there is little difference between the MD-55 and HS, since their active layers are of similar thickness with $32\,\mu$m and $40\,\mu$m, respectively. Since HD-810 has very low sensitivity, doses up to $20\,\text{kGy}$ could be resolved. This is about two orders of magnitude more than earlier calibration measurements for medical applications [249, 250, 255, 258].

It should be noted that independent from this work, but little later in time, a group at Lawrence Livermore National Laboratory has calibrated the RCFs for protons as well [263]. The reference has used a microdensitometer for the film analysis with different color filters, and different (and most likely as random as in this work) fit functions have been used. With a red color filter a film saturation starting at $OD = 2.5$ and with a blue filter a saturation of up to $OD = 5$ was found. This is expected since the red filter is close to the absorption peak of RCF (see fig. 3.5). The result from the blue filtered measurement is shown as dashed lines in figure 3.11 for a comparison with the calibration from this work. A very similar dose-dependence of the optical density was found, however, the numbers from the reference are not directly portable to the calibration of this work. The optical densities measured with the ArtixScan 1800f are measured with the wavelengths of the scanner lamp and are scanner-internally converted to grey scale. This is not directly comparable to the optical densities determined with a microdensitometer and requires a cross-calibration of both devices. It is concluded, that each scanner has to be independently calibrated for protons to obtain the most accurate results. In addition, the results in ref. [263] verify *a posteriori* the decision to use a transmission scanner in grey scale mode for maximum dose range and not to filter the white lamp for the red wavelength.

Up to date there is no theoretical description of the RCF response for high doses. Therefore, the measured energy deposition values are fit by a ninth-order polynomial with the fit parameters a_i and b_i listed in table 3.3:

$$E_{\text{dep}} = \exp\left(\sum_{i=0}^{8} a_i OD^{b_i}\right) \text{ in } \frac{\text{keV}}{\text{mm}^2}. \qquad (3.4)$$

The data points for low optical densities required very short irradiation times around one second. The beam was switched on and off by opening and closing a valve, therefore the minimum exposure time was about one second. The measurement error imposed by the unknown opening and closing times of the valve could be compensated by the use of the current digitizer. However, the beam current could not be decreased further without losing beam stability. Due to that, the measurement

	a_1	a_2	a_3	a_4	a_5	a_6	a_7	a_8	a_9
HD-810	17.7	-2.4	65.4	-244.6	494.5	-486.7	307.5	-140.2	13.9
MD-55	431.8	-817.9	1518.0	-3103.6	4001.8	-2784.9	1191.8	-447.2	35.5
HS	-855.2	1674.8	-2695.2	4872.3	-5535.9	3391.8	-1197.2	395.2	-25.8
	b_1	b_2	b_3	b_4	b_5	b_6	b_7	b_8	b_9
all	0.0	0.1	0.5	0.9	1.3	1.7	2.3	2.7	3.3

Tab. 3.3: Tabulated fit parameters for equation (3.4).

could not be performed for very low doses and the data had to be extrapolated for the MD-55 and HS for optical densities below $OD = 0.1$ [258]. The fit functions are

$$OD_{\text{extrapol.}}(\text{MD-55}) = 5 \times 10^7 + 3.74 \times 10^{10} \, OD^{1.19} \quad (3.5)$$

$$OD_{\text{extrapol.}}(\text{HS}) = 5 \times 10^7 + 2.6 \times 10^{10} \, OD^{1.19}. \quad (3.6)$$

The error of this calibration is very hard to estimate. The energy deposition error according to eq. (3.2) is the sum of errors from the temporally integrated charge measurement with the Faraday cup, the current digitizer, the discriminator and the counter, the error of the beam area and the error of the SRIM-calculation, respectively. The error of the charge determination is on the order of one percent. The beam had very sharp edges from the aperture, hence the area error can be neglected. The SRIM calculation has an error of 4 % [267]. Overall, the relative error according by propagation of uncertainty in the energy deposition is 5 %. Inhomogeneities in the beam profile resulted in inaccurate optical densities, with an error about $\Delta OD = 3\,\%$. The calibration error by the disagreement of the polynomial regression curve (the mean deviation) and the measured data are 3.3 % for the HD, 3.9 % for the MD-55 and 4.6 % for the HS, leading to a maximum error of about 5 %.

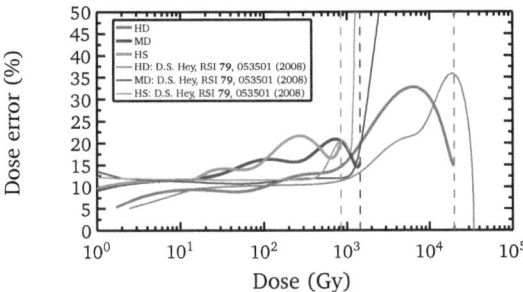

Fig. 3.12: Dose error in RCF calibration. The vertical dashed lines denote the dose limit corresponding to the practical optical density limit of $OD = 2.5$.

According to the manufacturer the production process produces batch-to-batch variations in the sensitivity of up to 10 % [263]. The calibration was performed with RCF from one batch only, however a comparison with data from earlier calibration attempts [252, 253] shows that the deviation might be below this value. The error in dose-determination (ΔD) imposed by the variance in the film sensitivity can be either calculated with eqs. (3.4) and (3.5) by replacing the optical density with

$OD' = (OD \pm 0.1\,OD)$ and determining the maximum deviation, or, equivalently, it can be obtained by the propagation of error as

$$\Delta D = \left|\frac{\partial \text{dose}}{\partial OD}\right| \Delta OD, \qquad (3.7)$$

with $\Delta OD = 0.1\,OD$. In both cases the dose is obtained from eq. (3.4) multiplied by $e(\rho d)^{-1}$. The result is shown in figure 3.12, with the thick lines being the calibration errors from this work and the thin ones from ref. [263]. Not unexpected, the curves from both works exhibit a very similar shape. The curves from this work show some oscillations, that are the result of the not-so-perfectly fitting ninth-order polynomials used.

The error in dose determination is on the order of 15 % up to doses of about 10 Gy. Then the error increases, that is due to the increasing slope in the data shown in figure 3.11, right. The curve for the HD (—) significantly increases for doses above 1 kGy, again due to the changing slope. For higher doses the films are close to the saturation of the scanner-RCF combination. At this point, little changes in optical density produce strong changes in the dose, hence the error in the dose determination increases. For optical densities above $OD = 2.5$ the films are saturated and the dose cannot be determined. The corresponding dose error would be 100 %. Since then an error estimate is useless, the plots are only shown up to the dose limits that correspond to the optical density of $OD = 2.5$, visualized as the dashed vertical lines. The limit is 20 kGy for the HD-810, 1.5 kGy for the MD-55 and 0.85 kGy for the HS, respectively.

In conclusion, this section presented the radiochromic film calibration for protons. The measured data have been fit with ninth-order polynomial functions, allowing to determine the energy deposited in the RCF by scanning the film for its optical density. The error of the dose-determination has been determined to be 20 % at maximum for doses below 1 kGy, and 30 % for doses above 1 kGy for the HD-810. The variance in the sensitivity from batch to batch during the production process has the strongest influence on the dose error for doses obtained in the proton-acceleration experiments.

3.4.3 Energy deposition and dose rate sensitivities

After the calibration, the curves have been tested for their prediction capability. For this measurement, a stack of RCF from a new batch has been placed in the beam line of the Tandem accelerator. The RCF stack consisted either of two layers of HD-810 followed by three layers of MD-55 or it consisted of four layers of HS. In front of the stacks, several aluminum foils with $12\,\mu$m thickness have been placed to decrease the projectile energy penetrating the RCF. The ray-tracing code described above has been used to calculate the Bragg-peak, determined to be sitting either in the middle of the three MDs or in the third HS. Similar RCF stack configurations are used in laser-proton-acceleration experiments for a determination of the proton energies by their range and for a determination of the particle number by the film coloration. Hence the stack used here should serve as a proof of the measurement technique. Each stack configuration has been irradiated for six different times, overall 50 measurements have been performed.

By comparing the expected film coloration with the measured one, it was found that near the Bragg-peak the film's optical density significantly deviates from the expected value, whereas the RCFs before show an optical density as expected. The reason is a dE/dx-sensitivity of the active layer, similar to the energy-deposition sensitivity found for x-ray films [268–270] or for RCF at low doses, but with unpublished stopping power [271]. The reason for less coloration with increasing stopping power is an increase of the *local* dose due to dE/dx as well as a decreasing track radius in a diacetylene monomer in the active layer, leading to a local saturation effect.

The measured data have been averaged over the six irradiation times. The resulting optical density has been compared to the expected optical density from eqs. (3.4) and (3.5). The ratio of both is the film detection efficiency [268]

$$\eta\left(\frac{dE}{dx}\right) = \frac{OD_{\text{measured}}}{OD_{\text{expected}}}. \tag{3.8}$$

This detection efficiency is then plotted versus the stopping power dE/dx, shown in figure 3.13 (left). Even though the measurement suffers from poor statistics, a clearly decreasing detection efficiency with increasing dE/dx is recognizable.

For large dE/dx values the efficiency decreases up to 50 %, cutting the measured optical density to half the expected value. Linear fits ($\eta = P_1 - P_2 \times dE/dx$) are shown for the MD-55 data (—) with $P_1 = 1$, $P_2 = 0.013$ and HS data ( ) with $P_1 = 1.1$, $P_2 = 0.017$, respectively. The black data points have been obtained from

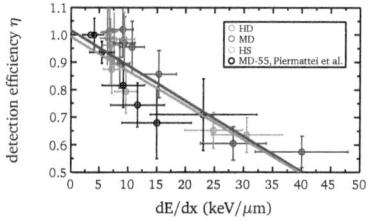 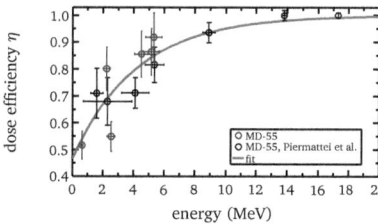

Fig. 3.13: Film coloration efficiency. The measured optical density has been compared to the expected optical density from eqs. (3.4) and (3.5). (left) The detection efficiency is plotted versus the stopping power dE/dx. (right) The detection efficiency of MD-55 versus the proton total energy. The black data points have been obtained from ref. [271] for MD-55.

relative depth dose measurements in MD-55 in ref. [271], figure 2, and a SRIM calculation of the stopping power performed in this work. These data have been included in the linear fit for the MD-55.

The calibration curves have been obtained at stopping powers of $6.1\,\text{keV}/\mu\text{m}$ for the HD-810, $6.6\,\text{keV}/\mu\text{m}$ for the MD-55 and $7.25\,\text{keV}/\mu\text{m}$ for the HS, respectively. For these values, the efficiency might already be less than one. Figure 3.13 (right) shows the efficiency from ref. [271], figure 2, and the values obtained in this thesis, multiplied by 0.9 as the data from ref. [271] suggest for 8 MeV protons. A fit with $\eta = 1 - \exp(-P_1 \times \text{energy} - P2)$ with $P_1 = 0.24$, $P_2 = 0.63$ shows reasonable agreement. This curve is used later on in this thesis to account for the detection efficiency.

The measurement errors are quite large. The error in the measured optical density is about 5 %, the calculated OD has an error up to $\Delta OD_\text{calc.} = \pm 15\,\%$. Hence the error in efficiency is 20 %. The error in the stopping power is mainly due to the uncertainty of the aluminum foil thickness ($\Delta x = 2\,\mu\text{m}$), resulting in 40% uncertainty in dE/dx. The data set given in ref. [271] is not complete as well, leading to large uncertainties in the stopping power up to 25 %. These large errors require a more sophisticated experiment to clarify the stopping power dependence of RCF coloration in a future work.

From the measurement above, the dependence of the RCF coloration on the dose rate can be determined as well. According to the literature [256], RCF are insensitive to the temporal duration of the applied dose, which has been determined for dose rates up to $4\,\text{Gy/min}$. In this work dose rates up to $1080\,\text{Gy/min}$ have been applied. Figure 3.14 (left) shows the measured optical densities in HS versus dose.

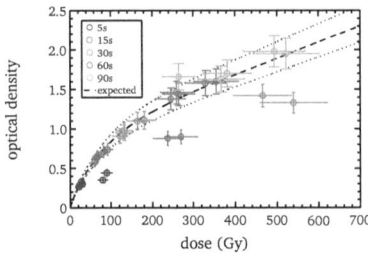
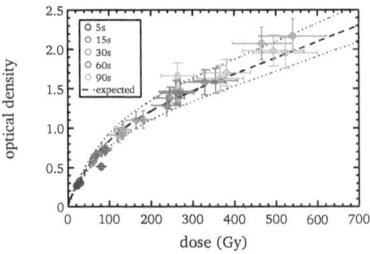

Fig. 3.14: Measured optical density versus applied dose (○) for different irradiation times. The data are compared to the expected optical density (- - -), under the assumption of ±10 % variation (· · ·). The left image shows the data as measured, the right image has been obtained by correcting the optical density with the detection efficiencies from figure 3.13 (—).

The irradiation times are color-coded and described in the figure legend. The data are compared to the expected optical density (- - -), with ±10 % variation (· · ·) according to the RCF optical density sensitivity from the manufacturer. Most data points are in the error interval, however some of them significantly deviate. These deviating values are only from these films that have been placed in the Bragg-peak, developing less OD due to the stopping power dependency. The right image of figure 3.14 shows the same plot, but with OD values corrected by the linear fit from figure 3.13 (—). A good agreement is now found, with most data points being in the error interval.

The RCF stack measurement has shown, that RCF suffers from a stopping power dependency of the film coloration. The current data imply that MD-55 and HS can develop up to 50 % less optical density than expected. The data for HD-810 are not sufficient for a statement. This stopping power dependence implies, that for a determination of an unknown particle number by measuring the optical density of the film this effect should be taken into account. Due to the poor statistics an error of 50 % is assumed in the measured dose.

Furthermore, RCF does not show a significant dependence on the dose rate up to 1 kGy/min with data acquisition times on a second level. This might change for time scales on the order of a nanosecond (the pulse duration of TNSA-protons), however these short times are not accessible with the current accelerator technology and would require an online calibration at a laser system.

3.5 RCF Imaging Spectroscopy

For the detection of laser-accelerated protons the calibrated RCFs are aligned in a stack. Figure 3.15(left) shows a schematic RCF configuration consisting of one layer of HD-810 and seven layers of MD-55 as an example. The TNSA-protons enter the stack from the left side. An RCF is sensitive to all ionizing radiation, but it is most sensitive to protons due to their higher stopping-power compared to electrons or x-rays. Heavy ions only penetrate the first layer. The whole RCF stack is wrapped in aluminum foil to further protect the RCFs from parasitic radiation as well as from target debris.

The protons penetrate the stack up to their range according to their energy. The energy-deposition in the active layers is calculated with the ray-tracing algorithm from above. For this calculation, the whole energy range to be calculated (here [0-15] MeV) is divided into 0.01 keV energy bins. Those energies are then used as initial energies. For each initial energy the code subsequently calculates the stopping in each RCF, taking into account the different compositions of the different types. The deposited energy in the active layers is then plotted versus the initial energy. This is shown in figure 3.15(right).

Each layer has a lower detection threshold where the energy deposition quickly rises and then slowly falls off, representing an "inverse" Bragg-curve. MD-55 has two maxima because of the two active layers in the film. The total energy deposition in MD-55 is higher than in HD-810 due to the thicker active layer. Because of this peaked energy deposition profile, each RCF layer can be attributed to a small energy interval with a width of 1 MeV for MD-55 and 0.5 MeV for HD-810, respectively. Hence a stack of RCF layers can be used as a two-dimensionally imaging spectrometer, measuring the transverse intensity distribution in two dimensions and the energy-resolved particle spectrum in the third dimension.

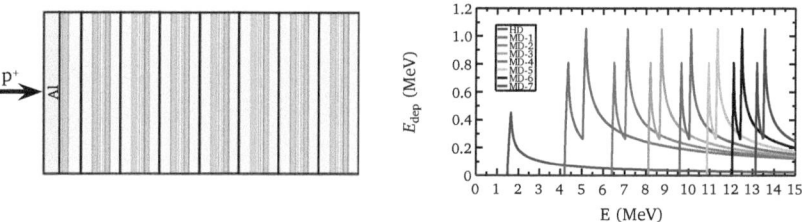

Fig. 3.15: (left) RCF stack alignment (1 HD-810, 7 MD-55) for laser-accelerated protons, (right) corresponding energy deposition in the RCF stack.

3.5.1 Opening angle

RCF measures the spatially resolved dose distribution of the proton beam. This is equivalent to a measurement of the angle of beam spread or opening angle of the proton flow, in a small energy interval according to the layer position in the stack. The opening angle represents momentum space information of the flow. This statement can be derived by the following procedure. It is assumed that the protons propagate ballistically (force-free). This assumption is justified as soon as the protons have propagated a distance of about a few 100 μm, as obtained from Particle-In-Cell (PIC) simulations and from a recent measurement [26]. The RCF-detector stack is usually a few centimeters away from the target, therefore the protons move on ballistic trajectories for most part of the path to the detector. The detector measures the distribution in the (x,y)–plane. Knowing the distribution at $t=0$ and by the assumption of ballistic expansion the position of the protons in the RCF at place $z = z_{\text{RCF}}$ can be calculated as follows:

$$x_i(t_i) = x_i(0) + v_{x,i} t_i, \quad y_i(t_i) = y_i(0) + v_{y,i} t_i, \quad z_{\text{RCF},i}(t_i) = z_i(0) + v_{z,i} t_i. \quad (3.9)$$

The times t_i are the individual arriving times for the individual protons i at the RCF, starting from different locations (x_i, y_i, z_i) at $t = 0$. Solving the last equation for t_i and inserting it into the other two results in

$$x_{\text{RCF}} = x_i(0) + \frac{v_{x,i}}{v_{z,i}}(z_{\text{RCF}} - z_i), \quad y_{\text{RCF}} = y_i(0) + \frac{v_{y,i}}{v_{z,i}}(z_{\text{RCF}} - z_i). \quad (3.10)$$

Since $z_{\text{RCF}} \gg z_i$, the essential information recorded in the RCF is the angular distribution of the flow. Writing $v_{x,y}/v_z = p_{x,y}/p_z \approx \vartheta_{x,y}$, the resulting equations are

$$x_{\text{RCF}} \approx x_i(0) + \vartheta_x\, z_{\text{RCF}}, \quad y_{\text{RCF}} \approx y_i(0) + \vartheta_y\, z_{\text{RCF}}. \quad (3.11)$$

The remaining variables are $x_i(0)$ and $y_i(0)$ - the initial transverse positions at the source. As an estimate, they are on the order of the laser focal spot size, i.e., about 10 μm. This is negligible compared to the cm-long propagation distance to the detector. Nevertheless, there is a unique method to directly measure the source size by imaging tiny grooves from the target rear side into the RCF, that will be described in the following section.

3.5.2 Micro-grooved target foils

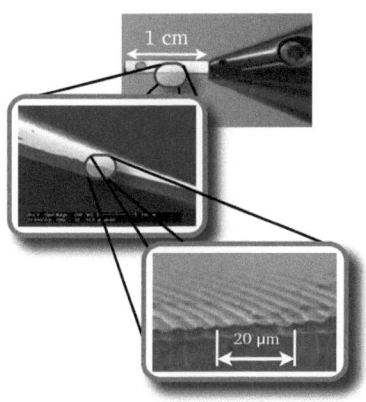

Fig. 3.16: Micro-structured, 50 μm thick gold foil. See text for details.

Soon after the discovery of laser-accelerated protons from the rear side, it was found that micro-corrugations on the target rear-surface lead to distortions in the proton beam, that are magnified and imaged into the RCF detector stack [54]. This unique property of TNSA-protons can be used to determine the real source size from where the protons originate [57, 72, 243]. Therefore, a periodic groove pattern is machined into the otherwise optically flat rear surface. The foils were made of gold due to its suitable mechanical properties. During this thesis, various micro-machining techniques have been tested. Good results have been obtained by direct diamond-planing of 50 μm thin gold foils. The diamond-planing of the foils has been carried out by the Institut für Mikroverfahrenstechnik at Forschungszentrum Karlsruhe [272]. (2×1) cm foil pieces have been first glued onto a mount and were then diamond-planed. After removing them from the mount they have been electropolished for a smoother surface. In a next step, the foil pieces were precision laser-cut [273] into thin stripes 2 mm wide and 1 cm long. An example is shown in figure 3.16. The photograph shows the target foil in its mount before an experiment. The hole on the left side has been machined for a different holder. The micro-grooves are not directly visible. The two gray-scale images have been obtained with a scanning electron microscope at GSI. The lower right image shows the wave-like groove structure at the rear side. The periodicity has been chosen to 5 μm peak-to-peak distance. The depth of the grooves is below 1 μm. Unfortunately, this direct diamond-planing is not applicable to foils thinner than 50 μm.

For arbitrary thin foils an indirect method has been developed. First, a brass or copper wafer is made by micro grinding. The mm-thick wafers with 75 mm diameter have been produced by the Laboratory for Precision Machining at Universität Bremen [274]. The further processing has been developed in the group's target laboratory at the TU Darmstadt in the framework of this thesis and a diploma thesis [252].

First, a layer of gold is deposited on the micro-structured surface by electro-plating. The thickness of the gold layer is adjusted by the current and time-duration while electro-plating [252]. In the next step, the wafer is removed by etching in nitric acid. The gold layer is unaffected by the nitric acid. Hence after this step a thin, micro-structured gold foil remains, that is then laser-cut to its target size. Foils of $5\,\mu$m, $10\,\mu$m, $15\,\mu$m and $30\,\mu$m thickness with $3.6\,\mu$m groove distance have been produced with this technique. The first samples had a structure that was just the inverse structure of that one shown in figure 3.16, later samples had a true sinusoidal structure [9].

The grooves are imprinted in the proton beam. Figure 3.17(left) schematically depicts the action of the grooves at the rear side, as already shown in the 2D-PIC simulation in fig. 2.17. The protons are being accelerated, with an initial direction normal to the local surface. Hence, the modulated surface leads to the so-called *micro-focusing* [72, 83] of the protons near the target surface[2]. The grooves act as tiny cylinder lenses, focusing the ions to individual beamlets about 200 nm wide [72]. For later times, the global sheath expansion leads to a transverse spreading of the whole beam. After the acceleration, the proton beam expands ballistically in free space, until it hits the RCF detector stack. According to their range, the protons deposit their energy in the individual layers of the stack, leading to a coloration. The micro-focused beamlets from the grooved surface locally lead to a higher energy deposition, hence the RCF becomes darker in this region. On the right side of figure 3.17 an RCF image of protons with 8 MeV energy is shown. The data have been obtained in an experimental campaign at the TRIDENT laser at Los Alamos National Laboratory, NM, USA. The laser delivered 20 J ($I = 3 \times 10^{19}\,\mathrm{W/cm^2}$) on the $10\,\mu$m thin gold target.

The grooved surface has led to lines imprinted in the RCF. By counting the lines and by multiplying with the known groove distance, the initial source positions $x_i(0)$ and $y_i(0)$ (see eq. (3.11)) can be obtained. Hence a correlation between the source position and the final opening angle can be obtained, leading to detailed information about the accelerating electron sheath [76, 87, 88]. By accelerating protons off non-periodic structures it was verified, that no overlap of the beam perturbations from adjacent structures appears [10]. The imaging of the surface requires a high

---

[2]T.E. Cowan, M. Roth and P. Audebert, "Method and apparatus for nanometer-scale focusing and patterning of ultra-low emittance, multi-MeV proton and ion beams from a laser ion diode", US Patent # 6852985, Feb. 8, 2005

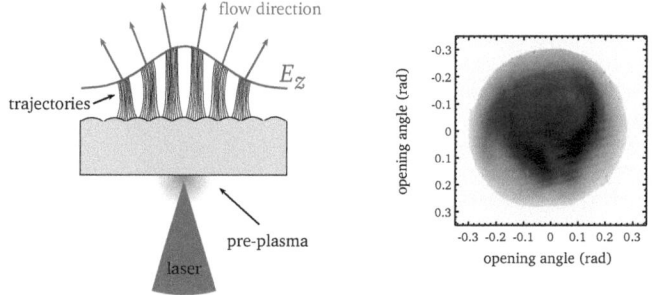

Fig. 3.17: Proton acceleration from a micro-structured foil.

beam laminarity, expressed by the ion beam *emittance*. The concept of the beam emittance, as well as emittance measurements of TNSA-accelerated protons will be described further in section 4.1.4. However, the image generation by the micro-grooves is not as simple as describe here and requires a description in the ion beam phase-space. First attempts have been made in refs. [87, 88]. Further investigations with PIC-simulations and a semi-empirical model have been made in the framework of this thesis. This will be the topic of chapter 5.

3.5.3 Spectral reconstruction of laser-accelerated protons

The RCF stack allows for a reconstruction of the particle number spectrum. It has to be taken into account, that high-energy protons that are being stopped in the layers at the end of the stack also deposit some of their energy in the layers before. The measured dose in one RCF is not only from protons having their Bragg peak in this layer, but it is the sum of all protons penetrating the RCF, weighted by the energy-dependent particle number. Thus, the spectrum has to be deconvolved from the energy-deposition in the whole RCF stack. For this task, a computer routine with graphical user interface (GUI) has been written in MATLAB. Figure 3.18 shows a screenshot of the program, running in MATLAB on Apple Mac OS X. It allows to comfortably and quickly analyze the scanned RCF stacks.

The routine requires an energy-deposition curve (see fig. 3.15) for each RCF in the stack. The RCF scans are first read into memory. Each layer is converted to optical density and the background from an un-irradiated part of the RCF is subtracted. Then the beam envelope has to be found to cut out the beam from the rest of the RCF. Since the beams are not perfectly round, this cannot be done by a simple circle

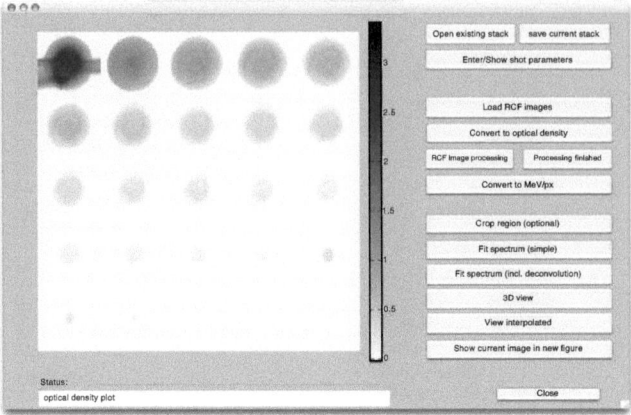

Fig. 3.18: Screenshot of the graphical user interface running in MATLAB on Apple Mac OS X for the reconstruction of the proton spectrum.

mask. MATLAB has powerful built-in image processing algorithms, that allow to find the beam envelope by a semi-automatic threshold detection. Furthermore, dust or scratches can be eliminated with the `roifill`-filter. The filter uses a polygon with an arbitrary number of nodes to encircle the scratch. `roifill` smoothly interpolates inward from the pixel values on the boundary of the polygon.

After this step, the optical density values of the RCFs are converted pixel-wise to the corresponding deposited energy in MeV/px. This is then summed to obtain the total energy deposited in each RCF layer. The energy E_{RCF}, deposited by the protons in each RCF, is the energy-dependent deposited energy per particle E_{loss}, convolved by the particle spectrum $\mathrm{d}N/\mathrm{d}E$:

$$E_{\text{RCF}} = \int_0^{E_{\max}} \frac{\mathrm{d}\,N(E')}{\mathrm{d}\,E} * E_{\text{loss}}(E')\,\mathrm{d}E'. \tag{3.12}$$

E_{loss} are the non-linear detector response functions, taken from the SRIM tables (see fig. 3.15). The integration is taken from 0 to $E_{\max}$, where $E_{\max}$ denotes the energy corresponding to the Bragg-peak in the last RCF layer recording a signal.

The spectrum $\mathrm{d}N/\mathrm{d}E$ could be determined by the inverse operation. This is numerically complex and potentially very sensitive to small perturbations in the initial

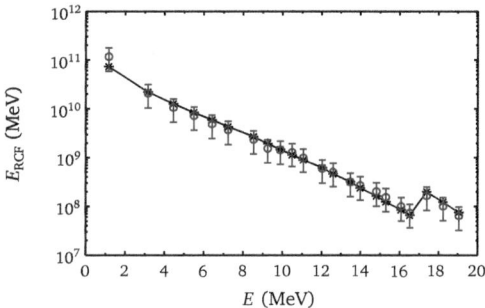

Fig. 3.19: Energy deposition E_{RCF} of TNSA-protons generated at the TRIDENT laser at Los Alamos National Laboratory. The maximum energy was 19 MeV. The measured energy deposition (○) could be fit by the calculated energy deposition (-∗-) assuming a spectrum according to eq. (3.14) with the parameters $N_0 = 3.5 \times 10^{11}$ and $k_B T = 8.74$ MeV.

conditions. Therefore a certain spectral shape is assumed and fit to the data. One of three potential spectra usually fits to the measured data. These are either a simple exponential decay

$$\frac{\mathrm{d}N}{\mathrm{d}E} = \frac{N_0}{E} \exp\left(-\frac{E}{k_B T}\right), \tag{3.13}$$

or a modified version of it

$$\frac{\mathrm{d}N}{\mathrm{d}E} = \frac{N_0}{E} \exp\left(-\frac{E^2}{(k_B T)^2}\right), \tag{3.14}$$

or the spectrum of a self-similar expansion from eq. (2.54)

$$\frac{\mathrm{d}N}{\mathrm{d}E} = \frac{N_0}{\sqrt{2k_B T E}} \exp\left(-\sqrt{\frac{2E}{k_B T}}\right). \tag{3.15}$$

The fit parameters N_0 and $k_B T$ are obtained by assuming some initial values and by calculating the deposited energy to be expected in the RCF layers according to eq. (3.12). This is then compared to the measured dose. A least-squares fitting algorithm then finds the best values for N_0 and $k_B T$. The fit parameter N_0 denotes the integrated particle number. $k_B T$ is not the proton beam temperature, but a mean energy. In the case of eq. (3.15) it is identical to the hot electron temperature driving the expansion.

Figure 3.19 shows as an example data obtained at the TRIDENT laser at Los Alamos National Laboratory. Protons from the rear side of a 10 μm thick gold foil could be accelerated up to 19 MeV energy. The proton beam profile is shown in the screenshot in fig. 3.18, with the energy increasing from top left to down right. The measured energy deposition (○) could be fit by the calculated energy deposition (–∗–) assuming a spectrum according to eq. (3.14) with the parameters $N_0 = 3.5 \times 10^{11}$ and $k_B T = 8.74$ MeV (∗). The last three films were MD-55 hence the energy deposition E_{RCF} was higher than in the HD-810's before. The resulting jump of the energy deposition is clearly visible, however the spectrum (eq. 3.14) remains smooth.

Chapter 4

Proton-acceleration experiments

This chapter is about the experiments performed at the four different lasers systems PHELIX, TRIDENT, LULI 100 TW and Z-Petawatt. In section 4.1, some typical parameters like the energy-spectra, angle of beam spread, source size and beam emittance are presented and compared to theoretical models. These results partially summarize the observations from a previous work [243]. In section 4.3 the influence of geometrical properties of the target foil on the control of accelerated protons is shown. This study is accomplished by the experiments presented in section 4.2, where the impact of the laser beam profile on TNSA-protons has been investigated. The chapter ends with the presentation of the results on the control and transport of laser-accelerated protons with the help of externally applied magnetic fields.

4.1 Typical parameters of TNSA-protons

The phrase "typical parameters" in the title of this section connotes that these parameters are the ones that are most important in the characterization of TNSA-protons. These are the energy-spectrum, i.e., the particle number per energy interval, the integrated number of protons, the maximum energy, the opening angle of the beam as well as the source size, respectively. The following subsections give an overview of the measured parameters at the four different laser systems and show their similarity, although the lasers vary in energy by one order of magnitude and in intensity by two orders of magnitude (see table 3.1).

4.1.1 Energy spectra of laser-accelerated protons

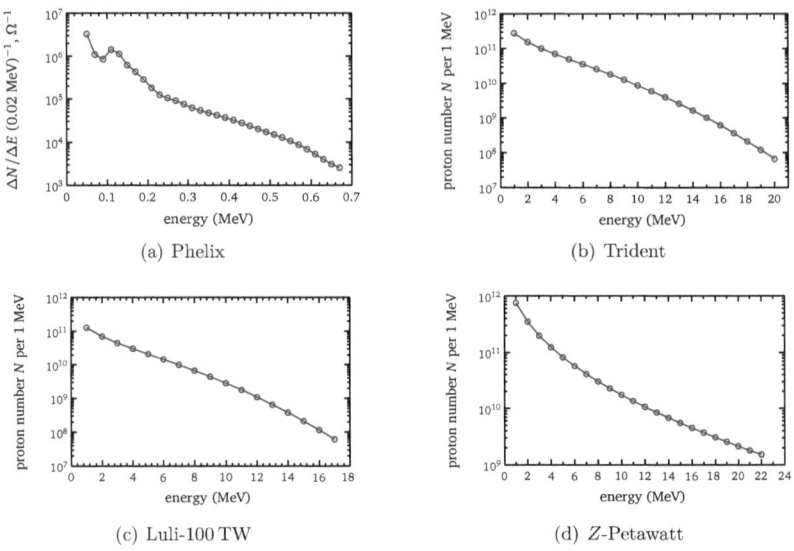

Fig. 4.1: Energy spectra obtained at a) PHELIX (Darmstadt), b) TRIDENT (Los Alamos, NM, USA) c) LULI-100 TW (Palaiseau, France) and Z-Petawatt (Albuquerque, NM, USA). The different laser parameters are given in the text.

Some exemplary energy-spectra are shown in figure 4.1. The spectrum shown in figure 4.1a was obtained at PHELIX at GSI in Darmstadt, Germany [129]. Since the laser has been still in its construction phase at the time of the experiments, only the preamplifier could be used. The laser pulse with a wavelength of $1.053\,\mu$m had an energy of 1.8 J on target, a pulse duration of 900 fs and could be focused to a spot with $11\,\mu$m full width at half maximum (FWHM). The energy in the focal spot was not measured, hence it is assumed that only 45 % are contained in the focus, similar to measurements at different laser systems [4, 275]. The intensity of the laser pulse was about $10^{18}\,\text{W/cm}^2$ ($a_L = 0.85$). The targets were $20\,\mu$m thin, micro-structured gold foils. The low energy of the accelerated protons did not allow to use a stack of RCF for their detection. Therefore, the spectrum was measured with a Thomson parabola ion spectrometer in a distance of 400 mm from the target. An aperture with 1 mm diameter has been placed at 225 mm distance to collimate the beam before it enters the spectrometer. A photograph of the target chamber interior is shown in fig. 3.1, the spectrometer itself is described in ref. [127].

The proton spectrum (see 4.1a) was obtained by integrating over four shots for a better signal-to-noise ratio. It shows a quasi-exponential decay up to a maximum proton energy of 0.7 MeV, where it ends with a sharp cut-off. The particle number is given per energy interval of 0.02 MeV and per solid angle ($\Omega = 1.5 \times 10^{-5}$ sr) of the Thomson spectrometer. The little enhancement in the particle number at 0.1 MeV could be due to the acceleration of heavy particles, that modify the accelerating field, leading to spectral collimation of the protons [90]. However, the statistics is not sufficient for a clear statement.

The spectrum shown in figure 4.1b has been measured at the TRIDENT shortpulse (C-beam) at Los Alamos National Laboratory, NM, USA [126]. The laser pulse parameters were: wavelength $\lambda_L = 1.053\,\mu$m, energy $E_L = 18.7$ J, pulse duration $\tau_p = 600$ fs and focal spot diameter $14\,\mu$m FWHM. About 45 % of the total energy are contained in the focal spot [4]. The laser irradiated the $10\,\mu$m thin, microstructured gold foil at a compound angle of 18.5° in elevation and 22.5° azimuthally to the target normal. The laser pulse with an intensity of $I_L = 9.1 \times 10^{18}$ W/cm^2 ($a_L = 2.5$) had an ASE contrast ratio of better than 10^{-6} at 1.2 ns before the main pulse.

The accelerated protons were detected with a stack of RCF, consisting of 19 HD-810 and nine MD-55 at a distance of (23±1) mm from the target. The stack had several brass layers in-between the RCF layers to further increase the energy range. The whole stack was wrapped in 12.5 μm thick aluminum foil to shield it from debris and parasitic radiation like electrons, x-rays or γ-rays.

After scanning and converting the optical density to deposited energy (eqs. (3.4) and (3.5)), the spectrum was determined by fitting one of the equations (3.13)-(3.15) to the measured dose. An optimum fit was obtained by equation (3.14) with $N_0 = 4.12 \times 10^{11}$ and $k_B T = 8.56$ MeV. As a measure for the quality of the fit the mean deviation between measured deposited energy and calculated values was determined to be of 15 %. A comparison between the measured deposited energy and the calculated one is shown in figure 3.19. The plot in figure 4.1b shows the particle number N per energy interval of 1 MeV, determined by piecewise integrating eq. (3.14). The protons could be measured up to the third MD-55, corresponding to a maximum energy of 19.5 MeV. The integration over the spectrum yields a conversion efficiency from laser energy to proton energy of $\eta = 1$ % for protons above 4 MeV. This threshold has been chosen, since protons accelerated by charge-separation at the front side of the target could be accelerated up to this energy [55, 62], hence it would obscure the conversion efficiency from laser to rear-side-accelerated protons.

Figure 4.1c shows a spectrum measured at the 100 TW laser at Laboratoire pour l'Utilisation des Lasers Intenses (LULI) at École Polytechnique, Palaiseau, France [128]. The laser had an energy of 18 J at 350 fs pulse duration, a wavelength of $\lambda_L = 1.057\,\mu$m and was focused to $8\,\mu$m FWHM. With 45 % energy in the focal spot, the intensity was $I_L = 3 \times 10^{19}\,\text{W/cm}^2$ ($a_L = 4.6$) impinging on the $15\,\mu$m thick, micro-structured gold target at normal incidence. The contrast ratio was 10^{-6} in 500 ps before the main pulse. The RCF stack in this experiment consisted of one layer of HD-810 and ten layers of MD-55, wrapped in $17.8\,\mu$m aluminum foil. The protons were detected at a distance of (45 ± 1) mm from the source. The corresponding RCF stack response function is given in figure 3.15.

The spectrum could be determined by fitting eq. (3.14) to the measured energy deposition, with the parameters $N_0 = 1.92 \times 10^{11}$ and $k_B T = 7.67$ MeV. The mean deviation between measured energy deposition and calculated one is 33 %. Since the laser intensity and target thickness are comparable to the TRIDENT, the spectrum is similar to figure 4.1b. Both particle number and maximum energy ($E_{\max} = 16.4$ MeV) are a little less than the TRIDENT results, due to the lower intensity and thicker target. The conversion efficiency for protons above 4 MeV is $\eta = 0.87\,\%$.

The last figure (fig. 4.1d) shows a spectrum from protons accelerated at the Z-Petawatt laser at Sandia National Laboratories, Albuquerque, NM, USA [130]. The data was obtained at the 100 TW target area, during the commissioning run of the PW-system. The CPA-laser at SNL with a wavelength of 1053 nm delivered 40 J laser energy on target, focused by the off-axis parabolic mirror to a beam spot of $5\,\mu$m full width at half maximum (FWHM). With a pulse duration less than 1 ps, the intensity on the target front side was $I > 9.2 \times 10^{19}\,\text{W/cm}^2$ ($a_L = 8.2$). Unfortunately, the pulse duration could not be measured during this experimental campaign. The pre-pulse contrast ratio was measured as 10^{-7}. A $25\,\mu$m thin Cu-foil was used as target, being hit by the p-polarized laser at an angle of 45°.

The RCF stack detecting the protons consisted of eight layers of HD-810, each one followed by two layers of $16.3\,\mu$m thick aluminum foil, and nine layers of MD-55 behind the HD-810 films. The aluminum foil wrapped around the stack was $16.3\,\mu$m thick as well. The stack was placed at (44 ± 1) mm behind the target.

In contrast to the spectra obtained at LULI and at TRIDENT, the proton spectrum measured at Z-Petawatt could be fit best with eq. (3.15), with the parameters $N_0 = 6.84 \times 10^{12}$ and $k_B T = 1.07$ MeV. The fit agrees up to 13 % mean deviation with the measured data. The particle number N_0 is higher than in the previous examples, as expected due to the higher laser intensity. The lower value for $k_B T$ is due to the different spectral shape used for the fit to the measured data, and not due to a lower hot electron temperature. The stack was not sufficient to resolve the maximum proton energy, since the scope of these experiments was not to determine the maximum energy. See section 4.4 for the details of these experiments. By comparing the cut-off particle numbers at TRIDENT and at LULI of about 10^7, it is concluded that the cut-off energy at Z-Petawatt was most likely around 30 MeV. The missing data at the end of the spectrum are most likely the reason for the difference in the spectrum function that best fits the measured values, since different shots with less laser energy have resulted in spectra with the shape of eq. (3.14) at this laser. Nevertheless, the integration of the spectrum delivered a conversion efficiency for protons above 4 MeV of $\eta > 1.05\,\%$.

The question arises, if the high particle number of more than 10^{12} protons could be supported by a contamination layer of hydrocarbons being present at the rear surface. This has been addressed by Allen et al. [55], who determined that $2.24 \times 10^{23}\,\text{atoms/cm}^3$ are at the rear surface of a gold foil, in a layer of $12\,\text{Å}$ thickness. Assuming an area of about $200\,\mu\text{m}$ diameter (see sec. 4.1.4), the accelerated volume is about $V = 3.8 \times 10^{-11}\,\text{cm}^{-3}$. Hence the total number of protons in this area is about $N_\text{total} = 8.4 \times 10^{12}$, that is close to the integrated number determined in the experiments.

4.1.2 Comparison with expansion models

The spectra shown in the previous section were shown as examples to depict the shape of the spectrum, the particle number as well as the maximum energy and energy conversion efficiency that can be obtained with 10-100 TW laser systems. More detailed studies on the dependence of the maximum energy and particle number, respective energy conversion efficiencies can be found in refs. [15–17, 127, 243]. References [15, 61] developed a scaling law at the LULI-100 TW to obtain the maximum proton energy and conversion efficiency in dependence on the laser parameters. The scaling law relies on an estimate of the laser-to-electron conversion efficiency and temperature, and a subsequent isothermal plasma expansion (see sec. 2.4.1) of the protons. The acceleration time has been fit as given in eq. (2.71). A comparison between the measured data and the values from the scaling law is shown in table 4.1. The scaling law reasonably agrees in maximum energy with the PHELIX, TRIDENT and LULI results, but it deviates for the conversion efficiency and for the

laser	I_L (W/cm^2)	E_{max} (MeV)	calc. E_{max} (MeV)	η (%)	calc. η (%)
PHELIX	1×10^{18}	0.7	0.8	-	-
TRIDENT	9.1×10^{18}	19	18.17	1	0.44
LULI-100 TW	3×10^{19}	16.4	18.12	0.87	1
Z-Petawatt	$> 9.2 \times 10^{19}$	≈ 30	44	1.3	2.6

Tab. 4.1: Comparison between the measured maximum energy E_{max} and conversion efficiency η versus calculated values from the scaling law described in sec. 2.4.1. The scaling law reasonably agrees in maximum energy with the PHELIX, TRIDENT and LULI results, but it deviates for the conversion efficiency and for the results obtained at Z-Petawatt.

results obtained at Z-Petawatt. The standard case in section 2.4.1 is very close to the TRIDENT and LULI experiments. For these parameters, that are similar to the laser parameters of the original publication [15], the model agrees well with the one-dimensional plasma expansion model. However, the scaling claimed in the reference predicts a maximum proton energy of about 44 MeV for the parameters of Z-Petawatt. The measured maximum energy of about 30 MeV is much smaller. This discrepancy for laser intensities above 10^{20} W/cm^2 has also been found by Robson et al. [16] at the VULCAN petawatt laser. The lower-than-expected energy is attributed to an acceleration in two phases with a first linear rise in the electron temperature while the laser is on, followed by an adiabatic expansion where the hot electron temperature quickly drops and further three-dimensional effects that cool the plasma even more.

Furthermore, it should be noted, that the scaling given in ref. [15] has been obtained with aluminum targets. The results presented here are from gold and copper foils, respectively. This could change the scaling relation, due to the different electron transport properties and surface contaminations of the different atomic species. In addition to that, it is not clear how different pre-pulse conditions (pulse duration, contrast level) affect the scaling. More systematic studies, that were not topic of this thesis, are required to clarify a possible influence of these parameters on the scaling relations.

Nevertheless, the shapes of the spectra are very similar for all laser-systems, supporting the model of a plasma expansion as shown in section 2.4.1. However, only in few cases the spectral shape resembles the shape of a self-similar, isothermal plasma expansion. In most cases, the particle number strongly drops for energies close to the cut-off energy. This kind of spectral shape has been measured at many laser systems [21, 56, 276–278]. Additionally, the result of the two-dimensional PSC-simulation (fig. 2.19) is of the same shape. Ref. [31] reports about a 3D simulation that fits this shape as well. The spectral shape from figures 4.1b,c and in the references can be fit by the relation given in eq. (3.14). Figure 4.2 shows the spectrum obtained in the 2D-PIC simulation (—), compared to best fits with eq. (3.13) (—), eq. (3.14) (—) and eq. (3.15) (—), respectively. The shape given by eq. (3.14) best fits the PSC spectrum for energies greater than 1 MeV. For $E > 8$ MeV the spectrum from the simulation is very noisy, since the quasi-particle number in the simulation was not sufficient to produce a smooth curve. The plateau in particle number below 1 MeV and the rise for even smaller energies is due to the existence of two electron temperatures [69], that are not taken into account in the fitting curves.

The shape from eq. (3.14) can be fit as well to the spectrum developing in a plasma expansion with charge-separation with adiabatic boundary conditions, as derived by Mora [69] (fig. 5). An adiabatic expansion is more realistic for later times, considering that the laser pulse is usually shorter than one picosecond. The acceleration time for protons to get e.g. 20 MeV kinetic energy at an electron temperature of 1 MeV (eq. (2.28)) and an electron density of 10^{20} cm^{-3} (eq. (2.34)) in the isothermal model is about 2 ps according to eq. (2.70). This is about twice the laser pulse duration. It is not very realistic to assume that the acceleration is fully isothermal for all times. The "real" acceleration process therefore is somewhere between the two extremes of an isothermal expansion and an adiabatic one.

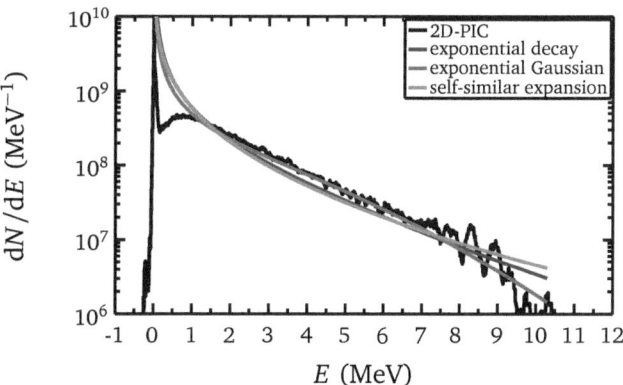

Fig. 4.2: Spectrum obtained in the 2D PSC-simulation from section A (—), compared to best fits with eq. (3.13) (—), eq. (3.14) (—) and eq. (3.15) (—), respectively.

4.1.3 Energy-resolved opening angle

The RCF Imaging Spectroscopy can be used to determine the energy-resolved opening angle of the proton beam, just by measuring the beam radius in each RCF. The radius in the RCF is equivalent to the angle ϑ, according to eqs. (3.11). Due to the stopping characteristics of protons in the stack, the signal in each layer can be attributed to protons having their Bragg-peak in this layer.

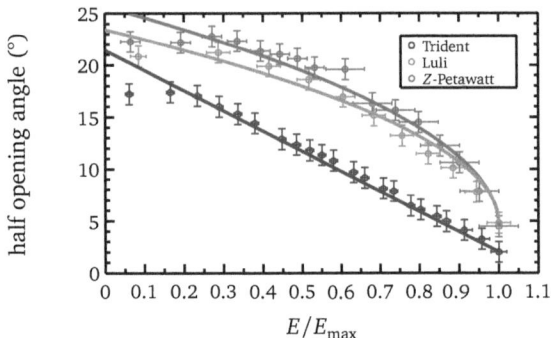

Fig. 4.3: Energy-dependence of the (half) opening angle. The data have been obtained at Trident (○), at LULI-100 TW (○) and at Z-Petawatt (○), respectively. The plots have been normalized to the respective maximum energy of each beam. The opening angle decreases with increasing energy. A parabolic dependency could be fit to the LULI and Z-Petawatt results, the data for Trident has a linear slope.

Figure 4.3 shows the energy-resolved opening angles for data obtained at Trident (○), at LULI-100 TW (○) and at Z-Petawatt (○), respectively. The plots have been normalized to the respective maximum energy of each beam. These are 19 MeV for TRIDENT, 16.3 MeV for LULI-100 TW and 20.3 MeV for Z-Petawatt, respectively. Protons with the highest energy are emitted with the smallest opening angle from the source, up to less than 5° half angle. Protons with less energy subsequently are emitted in larger opening angles. Below about 30 % maximum energy, the opening angle reaches a maximum and stays constant for lower energies. In most cases, the opening angles decrease parabolically with increasing energy, indicated by the parabolic fits to guide the eye. In some shots, however, the decrease of the opening angle with increasing energy is close to linear as in the example obtained at TRIDENT. The slope of the opening angle with energy is a result of the initial hot electron sheath shape at the target surface, as pointed out by Carrol *et al.* [106]. According to the reference, a sheath with Gaussian dependence in transverse direction results in a strongly curved opening angle-energy distribution, whereas a parabolic hot electron sheath results in a linear dependency. However, only crude details about the exact modeling of the acceleration process are given in the reference.

In the framework of this thesis, a more detailed expansion model has been developed, that is able to explain the experimental results in more detail. The model and its results are described in section 5.2.

It should be noted, that the term "opening angle" is not equivalent to the beam "divergence". As will be shown below, the divergence of the protons slightly *increases* with increasing energy, whereas the emitting area (source size) *decreases* with proton energy [72, 279]. This results in a total decrease of the opening angle measured with RCF.

4.1.4 Energy-resolved source sizes

Another feature that can be used in RCF Imaging Spectroscopy (RIS) is the ability of the proton beam to image corrugations of the target surface in the detector (see sec. 3.5.2). The grooves imprint in the beam and appear as darker lines in the detector. Due to the large opening angle of the beam, the grooves are magnified up to a factor of 1000. By counting the lines in each RCF and by multiplying the number with the known grooves distance, the source size at the target rear side can be determined. When this is done for each RCF layer, the source size is obtained with energy resolution.

Figure 4.4 shows energy-resolved source sizes for the three laser systems TRIDENT (∘), LULI-100 TW (∘) and Z-Petawatt (∘), respectively. As in the section before, the energy axis has been normalized to the individual maximum energy of the shot, with the maximum energies given in the section before. The source size decreases with increasing energy. Protons with the highest energies are emitted from sources of about 10 μm diameter and less. For lower energies, the source sizes progressively increase, up to about 200 μm diameter for the lowest energies measurable with RIS, that are about 1.5 MeV. For even lower energies, the source sizes might be much larger and could reach more than 0.5 mm in diameter [75].

The energy-dependence of the source size well fits a Gaussian, indicated by the lines in figure 4.4. The data could be fit by

$$E = \exp\left(\frac{-(4\ln(2)\,\text{source size})^2}{\sigma^2}\right), \tag{4.1}$$

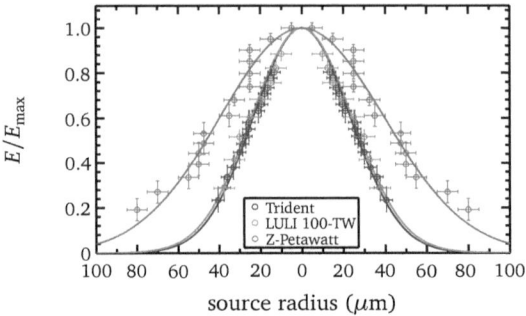

Fig. 4.4: Energy-resolved source sizes for data from TRIDENT (○), LULI-100 TW (○) and Z-Petawatt (○), respectively. The energy-source size distribution could be fit to a Lorentzian (eq. (4.1)) with FWHM $\sigma = 54.8\,\mu$m for TRIDENT, of $\sigma = 56.5\,\mu$m for LULI-100 TW and of $\sigma = 92.8\,\mu$m for Z-Petawatt, respectively.

where 2σ denotes the full width at half maximum (FWHM). This fit allows to characterize the complete energy-dependent source size with one parameter only. The FWHM for Trident with a $10\,\mu$m thin gold target is $\sigma = 54.8\,\mu$m. For LULI-100 TW the source size is $\sigma = 56.5\,\mu$m for a $15\,\mu$m thin gold foil. A larger source size has been measured at Z-Petawatt with $\sigma = 92.8\,\mu$m and a $25\,\mu$m thick gold target.

Decreasing source sizes have also been measured in refs. [72, 74, 75, 77, 119], but in most cases with less resolution and no statements are made for the shape of the energy-source size distribution.

The energy-dependence of the source size is directly related to the electric field strength distribution of the accelerating hot electron sheath at the source. Protons with high energies have been accelerated by a high electric field. Cowan *et al.* [72] relate an increasing source size with decreasing energy to the shape of the hot electron sheath, under the assumption of an isothermal, quasi-neutral plasma expansion (see sec. 2.4.1) where the electric field is $\boldsymbol{E} = -(k_B T_e/e)(\nabla n_e/n_e)$ (eq. (2.49)). A transversally Gaussian electric field distribution (as measured with RIS here) would result in a non-analytic expression for the electron density n_e. On the other hand, the realistic assumption of a Gaussian hot electron distribution would result in a radially linearly *increasing* electric field, in contradiction to the measured data. Hence it is concluded that the quasi-neutral plasma expansion, even though being the driving acceleration mechanism for late times, is not the physical mechanism explaining the observed source sizes.

In fact, the source size must develop earlier in the acceleration process, e.g. at very early times when the electric field is governed by the Poisson equation (eq. (2.35)), with $E(z) \propto k_B T_e/\lambda_D \propto \sqrt{k_B T_e n_e}$ (eq. (2.38)). With the data from figure 4.4 it is concluded, that there must be a radial dependency of $E(z)$, hence a Gaussian decay of either the hot electron temperature or density or both.

4.1.5 Beam emittance

An important parameter in accelerator physics is the transverse emittance of an ion beam. In view of the nature of the ion sources used in conventional accelerators, there is always a spread in kinetic energy and velocity in a particle beam. Each point on the surface of the source emits protons with different initial magnitude and direction of the velocity vector. The *emittance* ε provides a figure of merit for describing the quality of the beam, i.e., its laminarity [280]. The results presented in this section were partly taken from [252]. More results obtained with RCF Imaging Spectroscopy will be published in ref. [9].

Assuming that the beam propagates in the z-direction, the transverse emittance $\varepsilon_{x,y}$ can be determined with the help of the imaged micro-structures from the target rear side. As shown above (eq. (3.11)), each point on the RCF represents a point in the position-momentum space $(x, p_x$ and $y, p_y)$, the *phase space*. The transverse phase space (e.g. in x-direction) of the TNSA-protons is obtained by plotting the source position (indicated by the imprinted surface groove in the RCF) versus the angle of emission p_x/p_z, obtained by the position x of the imprinted line in the RCF and the distance d by $p_x/p_z = x' = \arctan(x/d)$.

Figure 4.5 shows the transverse phase space

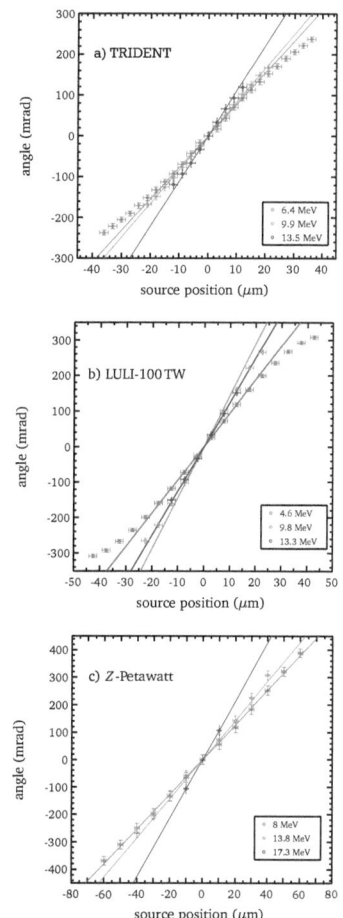

Fig. 4.5: Energy-resolved transverse phase space.

plots obtained at the three laser systems TRIDENT, LULI-100 TW [252] and Z-Petawatt. For each data set, the angle-position relation is given for three energies mentioned in the figure legends. The angle-position relation nearly linear, with increasing values for increasing distance from the center. The increasing angle with increasing initial position just reflects a diverging beam; a parallel beam would have zero angle for each position and a converging beam would have decreasing angles with increasing position, i.e., the line would be rotated by 90°. A perfectly laminar beam would resemble a straight line with zero width in transverse phase space [280]. This is indicated by the straight lines in the plots. Close to the center the angle-position distribution indeed shows a very linear slope. The data points for extended positions deviate from the straight line, since the lines in the RCF are closer to each other at the beam edge than in the center. This means that the angle growth $x'_{i+1} - x'_i$ decreases, indicating that the beam is not perfectly diverging. This is well pronounced for the 4.6 MeV data points in figure 4.5b. A bending effect has also been found in other experiments and computer simulations [72, 86]. The reason for the bending is the hot electron sheath that has accelerated the protons. It acts close to an ideal defocusing ion lens, but it has aberrations for protons at the outer positions of the beam.

The phase space representation of the protons rotates counter-clockwise with increasing energy. This means, that protons with higher energy (or momentum p_z), originating from the same transverse initial position, have a larger emission angle x' and therefore a larger transverse momentum, since $x' = p_x/p_z$. Hence protons with higher energies have a larger *divergence* than lower energetic ones. At the same time, the source size decreases proportional to $\exp(-x^2)$, therefore protons with higher energy are emitted with smaller opening angles. Computer simulations support the counter-clockwise rotation in phase space [86] and higher divergence [52] of higher-energy protons.

A perfectly laminar beam of charged particles is displayed by a line of vanishing thickness in the transverse phase space. Laser-accelerated protons are non-laminar particle beams, because they have a transverse velocity component deviating from the straight line, so the line in phase space is broadened. This deviation from laminarity is described by the emittance, describing the expansion or the parallelism of single particle trajectories. The definition of the transverse emittance $\varepsilon_{\text{trans}}$ in the phase space is the area of an ellipse including the proton beam, divided by π:

$$\varepsilon_{\text{trans}} = \frac{1}{\pi} \iint dx dx', \qquad (4.2)$$

in units of π mm mrad. Another definition of the emittance is the root-mean-square of the moments in the xx'-space:

$$\varepsilon_{\text{trans,rms}} = \sqrt{\langle x^2 \rangle \langle x'^2 \rangle - \langle xx' \rangle^2}. \qquad (4.3)$$

The term $\langle xx' \rangle^2$ reflects a correlation between x and x', that occurs when the beam is either converging or diverging. The relation between both quantities is $\varepsilon_{\text{trans}} = 4 \times \varepsilon_{\text{trans,rms}}$.

The emittance, as defined here, has to be modified to describe the quality of the beam because it depends on the kinetic energy of the protons. According to Liouville's theorem, the emittance does not remain constant with changing proton energy. The energy change is inversely proportional to the relativistic parameters $\gamma = 1 + E_{\text{kin}}/m_p c^2$ and $\beta = (1 - \gamma^{-2})^{1/2}$. Geometrically this corresponds to an increase or decrease of the slope x' (and hence the area in xx'-space) as the longitudinal momentum p_z is changed. Therefore the *normalized emittance* is introduced to compensate it:

$$\varepsilon_{\text{trans,norm}} = \beta \gamma \varepsilon_{\text{trans}}. \qquad (4.4)$$

Unfortunately, there is no global definition of emittance that is consistently used in accelerator- and ion beam physics. The emittance is sometimes defined as a phase space area, divided by π or not, or by using the root-mean-square emittance. Since the RIS data contains not enough information about the proton distribution to calculate the moments, the integration of the ellipse in phase space is used in this work. The bending of the phase space representation artificially decreases the emittance [86], therefore only 70 % of the whole data are used.

The normalized emittance was determined by calculating the semi-minor and semi-major axes of the ellipse fitted to the data in the transverse phase space. For the data obtained at TRIDENT, the normalized emittance decreases for increasing proton energy from $\varepsilon_{\text{trans,norm}} = 0.07\,\pi$ mm mrad to $0.01\,\pi$ mm mrad. For the LULI-100 TW it decreases from $\varepsilon_{\text{trans,norm}} = 0.13\,\pi$ mm mrad to $0.01\,\pi$ mm mrad and for Z-Petawatt it decreases from $\varepsilon_{\text{trans,norm}} = 0.14\,\pi$ mm mrad to $0.06\,\pi$ mm mrad, respectively. Conventional compact proton accelerators, e.g. the new HIT - Heidelberger Ionenstrahl-Therapiezentrum at Heidelberg, Germany operate at a normalized emittance of about 5π mm mrad [85]. Compared to that, the transverse emittance of laser accelerated proton beams is up to two orders of magnitude lower. The low emittance and therefore high laminarity allows for a strong focusing of the beam to small beam diameters. An estimate on the minimum diameter that can be obtained by focusing TNSA-protons is to trace back the proton trajectories from the RCF up to the point, where all ions seem to originate. This *virtual source* is usually several ten micrometers in front of the target [61, 74] and has about $10\,\mu$m diameter and less [9].

4.2 Laser beam-profile impression on laser-accelerated protons

The results from above were obtained with a round laser spot, focused as good as possible to obtain the highest intensities. But, as found by Fuchs *et al.* [76], the laser focal spot shape eventually imprints in the accelerated proton beam. The authors assumed that the bulk of the hot electrons follows the laser focal spot topology and creates a sheath with the same topology at the rear side. The proton beam spatial profile as detected by a film detector was simulated with a simple electrostatic model. The authors took the laser beam profile as input parameter and assumed the electron transport to be homogeneous, with a characteristic opening angle that needed to be fit to match the measured data. The unknown source size of the protons was fit to best match the experimental results. It was shown, that for their specific target thickness and laser parameters, the fitted broadening angle of the electron sheath at the rear side closely matches the broadening angle expected by multiple Coulomb small-angle scattering. However, they could only fit the most intense part of the measured beam and have neglected the lower intense part that originates from rear-side accelerated protons as well. Additionally, there is no information on the dependence of these findings on target thickness.

This section presents further studies that concentrated on the influence of the transverse laser beam profile on laser-accelerated protons. The laser beam profile was stepwise changed from the round, best focused spot to an astigmatic line focus. In contrast to the experiments described in ref. [76], the source size was simultaneously measured by using the micro-grooved targets.

The experiments were performed at the TRIDENT and the LULI-100 TW laser systems. The 30 TW beam-line of TRIDENT delivered 25 J in 800 fs on target. The

$f/3$ off-axis parabolic mirror focused the pulse to a FWHM focal spot of $12\,\mu$m, leading to an intensity above 10^{19} W/cm^2. The focal spot was measured with a high resolution, windowless Active Pixel Sensor (APS) in CMOS technology without any optical imaging to avoid misinterpretation of the beam profile due to aberrations. The pixel size of the detector was $3.5\,\mu$m with zero inter-pixel spacing and allowed to resolve the best focus. The position of the APS surface was accurately controlled with a commercial interferometer to be in the plane of the target surface.

The LULI-100 TW laser pulse was focused at normal incidence with an $f/3$ off-axis parabolic mirror to a FWHM focal spot of 6 μm, that lead to intensities above 5×10^{19} W/cm^2. The image of the focus was recorded with a microscope system and a CCD-camera. In both experiments the laser focus was either best focused or – to measure the influence of its beam profile on the proton beam profile – deformed to an elongated ellipse by tilting the parabolic mirror. The targets were $50\,\mu$m gold foils with chamfered, equidistant grooves with a line-spacing of $5\,\mu$m at the rear side. The accelerated protons were detected with an RCF stack, placed behind the target at either (23.5 ± 1) mm at TRIDENT or (42 ± 1) mm at LULI.

First, the results obtained with a round, best focused laser-beam spot are shown. The laser system was the LULI-100 TW. With a nearly diffraction limited laser focal spot, shown in fig. 4.6a, with FWHM= $6\,\mu$m, laser energy $E_L = 15.4$ J, intensity $I = 5.6 \times 10^{19}$ W/cm^2 and a target thickness of $50\,\mu$m with $5\,\mu$m grooves at the rear side, the resulting proton beam is round and smooth; see Fig. 4.6b. This is a result of the uniform laser focus as well as a smooth electron transport, first demonstrated by Fuchs et al. [76]. The decrease of the source size with energy, normalized to its maximum energy of 16 MeV, exhibits a gaussian distribution with a FWHM of $60\,\mu$m. These data are comparable to results published in ref. [72].

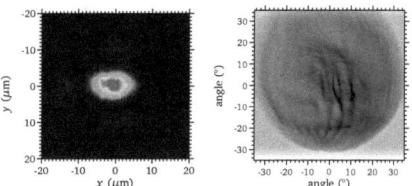

Fig. 4.6: (a) Round laser focus profile (LULI #25, FWHM= $6\,\mu$m, $E_L = 15.4$ J) that was used to irradiate a $50\,\mu$m Au foil. Its rear sided grooves with $5\,\mu$m distance could be imaged into the RCF that was placed at 42 mm distance. (b) The resulting proton beam profile (5 MeV) is round, the source size is $75\,\mu$m.

Next the laser focus was subsequently changed from best focus to a line focus, while the target thickness was kept unchanged. When the laser was defocused to form an elliptical line (fig. 4.7b-d), the proton beam profile followed (images show 5 MeV protons). The most intense part of the beam in fig. 4.7d roughly resembles an ellipse like the laser beam. Note that the orientation of the proton beam ellipse is perpendicular to the laser beam ellipse. This can be understood by following the argument in ref. [76]: the divergence angle of the protons is largest where the gradient of the electron sheath is strongest, that in turn has a similar shape as the laser beam profile. Therefore, the transverse acceleration is strongest along the short half-axis of the sheath ellipse and weaker along the long half-axis.

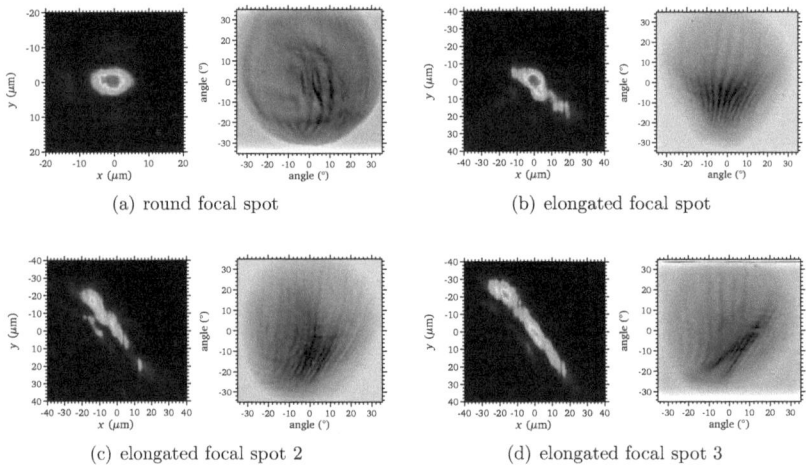

(a) round focal spot (b) elongated focal spot

(c) elongated focal spot 2 (d) elongated focal spot 3

Fig. 4.7: The laser was subsequently changed from a round focal spot (a) to an increasing ellipse (b-d). At its final stage (d), with $(80 \times 10)\,\mu$m axes length, the resulting proton beam profile shows the laser imprinted ellipse and a relatively round beam with lines. The source size is $75\,\mu$m, the size of the elliptical region is $35\,\mu$m in one direction.

The elliptical part (in one direction) of the 5 MeV protons in fig. 4.7d has four to seven lines, depending on which part of the dark area is assigned to the ellipse. The corresponding source size is then $(20 - 35)\,\mu$m, which is very close to the size of the laser focus. Note that even the lower intensity part of the beam shows lines. The visibility of the lines demonstrates that these protons originate from the rear side and have to be considered in an analysis, in contrast to ref. [76]. Additionally, the full-beam envelope is relatively round. The source size of the full beam is deduced to be $(85 \pm 5)\,\mu$m for these 5 MeV protons. This analysis was done for all RCF layers to obtain the full-beam source-size dependent energy distribution. It can again be fitted to a Gaussian as in the case with a round laser focus. The FWHM of the source is 92 μm, that is larger than the source size with a round focus. The targets were made out of the same foil, so the increase in source size can only be a result of the larger laser focus. Larger absolute values of the source size were expected due to the larger laser focus. However, the round envelope, similar angle of beam-spread and "Gaussian-like" decrease with increasing energy as in the case with round foci shows, that the electron sheath envelope and acceleration was the same in both cases, independent of the laser beam profile. The laser beam profile nevertheless had an influence on the spatial beam profile, but it seems to have the character of an initial perturbation, bending the lines inside due to a central non-radially symmetric part in a radially symmetric sheath.

For a further understanding of proton acceleration with an elongated laser spot the Sheath-Accelerated Beam Ray-tracing for IoN Analysis code (SABRINA) has been developed. The code and the results will be presented in section 5.1. The results have been published in ref. [2].

4.3 Beam optimization by target geometry

The accelerated protons can be shaped on a nanometer-scale by micro-machining the target rear surface. However, this does not have a significant influence on the accelerating electron sheath. For an optimization of the TNSA-process, i.e., achieving higher energy and better conversion efficiency, the conversion of laser light to hot electrons must be increased. This can be done just by increasing the laser intensity (see. eq. (2.21)). However, most laser systems already operate at their peak power and there is no simple way of further increasing the intensity. Another possibility is the modification of the geometry of the laser-plasma interaction zone at the target front side, e.g. by confining the pre-plasma in a cone geometry as described below.

4.3.1 Cone-shaped target front side

Cone targets are of interest for their potential to increase the hot electron temperature and population density [62, 281], which are the main contributors to the efficacy of the TNSA mechanism. Sentoku *et al.* [62] showed that sharp tip cones can effectively increase the number of electrons available for laser heating while guiding the laser light along the cone wall surface toward the cone apex. This action tremendously increases the interaction area of the laser, producing more electrons and concentrating the laser field at the cone neck near the flat-top surface. Special cone targets have been produced by T.E. Cowan and co-workers from University of Nevada, Reno (USA) and the nanofabrication group NanoLabz [282]. The cones were agreed to be shot at the TRIDENT by T.E. Cowan and B. M. Hegelich under the experimental leadership of K.A. Flippo. The author has contributed to the experiment by analyzing the RIS data.

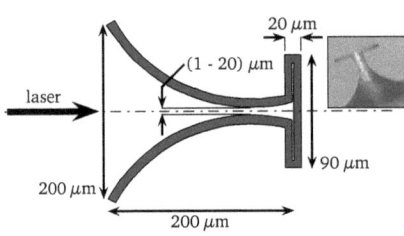

Fig. 4.8: Cone-on-flat foil target schematic, for explanations see text. Image courtesy of K. Flippo.

The laser parameters have already been given in section 4.1.1. The targets consisted of flat, microstructured gold foils and flat-top cone targets with a novel curved open-end design stemming from the lithographic technique from which they are produced. Cones with a range of neck and flat-top diameters were used in this experiment. The cone targets, schematically shown in figure 4.8, were prepared and mounted on glass stalks such that the flat-tops would be aligned normal to the incoming laser's compound angle (zero-degree incidence). Five targets were mounted so that the flat-top of the cone was not aligned normal to the laser path, but aligned only azimuthally, such that the laser would enter the cone off-axis in elevation at an incidence angle of 18.5°. The proton beams were diagnosed using RIS. All three types of RCF were used: HD-810, MD-55 and HS. Tastrak® CR-39 nuclear track detector was also used to record the proton beam. Unlike RCF, which is self developing, the CR-39 needs to be developed in a six-molar solution of NaOH for one hour to reveal the tracks left by the protons. Care must be taken not to overdevelop the CR-39 as it can lead to artifacts [283]. CR-39 is insensitive to electrons, x-rays, and gammas and has been used since the first forward directed short-pulse ion experiments [18, 19, 22] as an ion detector in short-pulse matter interactions.

Several 2 mm wide, 15 mm tall and 10 micron thick gold flat-foils were shot using the TRIDENT system as a base-line; these foils yielded a maximum proton cut-off of (19 ± 1) MeV for a laser energy of 18.7 J. The energy spectrum is shown in fig. 4.1b. The RCF stack from a $10\,\mu m$ thick gold cone with a neck outer-diameter of $25\,\mu m$ and a flat-top of $100\,\mu m$ is shown in figure 4.9. The stack was again wrapped in

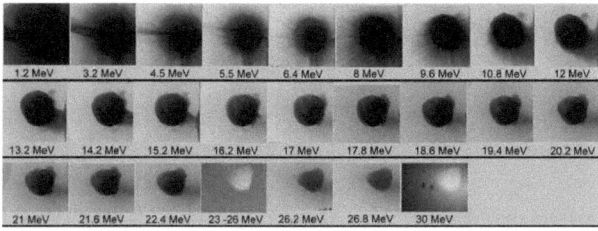

Fig. 4.9: Cone target with a top diameter of 100 μm and a neck of 25 μm shot at an intensity of $I_L = 10^{19}$ W/cm^2. The stack consists of 5 HD-810, 9 MD-55, 7 HS, a piece of CR-39, 2 HS, and a piece of LANEX regular on the back.

13 μm of Al foil, but consisted of multiple layers of HD-810, MD-55, HS, with a 1100 μm thick piece of CR-39, followed by two more pieces of HS. On the back of the stack, outside of the Al wrapping, a piece of Kodak LANEX® regular was attached originally planned to monitor the electron emission from the target; however, unexpectedly, it revealed the proton beam exiting the stack in excess of 30 MeV. The proton beam was also recorded on the CR-39's front and back surfaces as it passed through the plastic. Looking at the last few RCFs in the stack and comparing the shape and size of the beam on the LANEX, the CR-39 and the RCF, it can be determined that the beam exiting the rear of the stack is indeed protons. The signal cannot be due to heavier ions as they would not have penetrated the stack to such a distance, nor electrons, x-rays, or γ-rays as they would not have left tracks in CR-39. This particular cone, which had the highest performance in terms of the maximum proton energy, had a top to neck diameter ratio of four.

Many cones of various top and neck diameters were shot during these experiments. There is no obvious conclusion on whether top or neck cone diameter is correlated with proton energy. Nevertheless, as the size of the flat-top and neck increase, it becomes more and more like the flat-foil case. Using the RIS data to plot the proton energy as a function of proton number, shown in figure 4.10, one can easily see that the flat-top cone (– ○ –) produces more protons at higher energies than the flat foil (– ○ –), in this

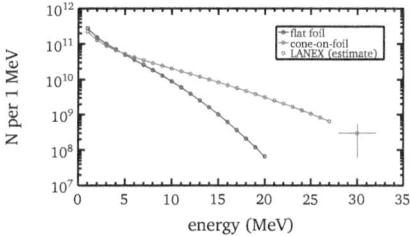

Fig. 4.10: Cone (– ○ –) versus flat-foil (– ○ –) spectra.

Chapter 4. Proton-acceleration experiments

case 13 times more above 10 MeV. The reconstructed proton spectra for the cone and the flat-foil targets can be summed to find the total energy conversion, again for protons above 4 MeV energy. The total energy present in the beam from the Au flat-foil is measured to be 187 mJ, or $\eta = 1\%$ of the incident 18.7 J laser energy. When this is done for the flat-top cone target it is measured to be 532 mJ, corresponding to 2.8 % of the 19 J of incident laser energy. This represents a nearly 3-fold increase in the conversion efficiency over the Au flat foil targets and a 1.5-fold increase in the total amount of protons, with nearly 5 times the number above 10 MeV. The beam also has potentially more than twice the maximum energy as simulations indicate [4]. Compared to other similar glass laser systems, the proton beam observed from the cone target contains two orders of magnitude more protons in a 1 MeV bin at 10 MeV than initial experiments at similar intensities on the CUOS T-cubed laser [236] and an order of magnitude more than recent experiments reported from the LULI-100 TW and RAL VULCAN lasers for the same intensity [15, 279]. The flat-top cone beam is more than two orders of magnitude more efficient (laser-energy conversion to protons) than previously published laser data and scalings for the same intensity on the LULI-100 TW [15], and nearly twice as high for the same laser energy on the RAL VULCAN laser [16], though similar or higher efficiencies have been reported at significantly higher laser energies and intensities [16, 279]. The flat-top cone proton beam also has 3-10 times higher maximum proton energy than previously reported for a similar intensity from experiments [15, 16] and scalings [15–17] of flat foils, with only the RAL VULCAN [16] and NOVA Petawatt [21] reporting higher proton energies, but again for much higher laser irradiances and energies. These comparisons should be taken with the caveat that laser systems are difficult to compare due to specific laser and target performance issues such as laser ASE contrast, system pre-pulses, pulse compression, pulse duration, B-integral, near-field beam quality, focusing optics, focal quality, as well as target thickness, material and surface quality, just to name a few.

It is apparent from figure 4.9 that the maximum proton energy from the cone was not captured on the RCF or even the LANEX screen as the beam size and the number of ions is still rather large, indicating that it is still far from the cut-off energy (as compared for example to the flat-foil target in figure 3.18). Figure 4.11 shows two contour plots of proton energy vs. 2-D divergence vs. proton number from the RIS data. The y-axis is the normalized proton energy, $E_{\text{proton}}/E_{\text{max}}$, where E_{max} is the cutoff energy observed in the RCF stack, the x-axis is the 2-D angular beam divergence θ, and the proton number in decades is indicated by the color shown in

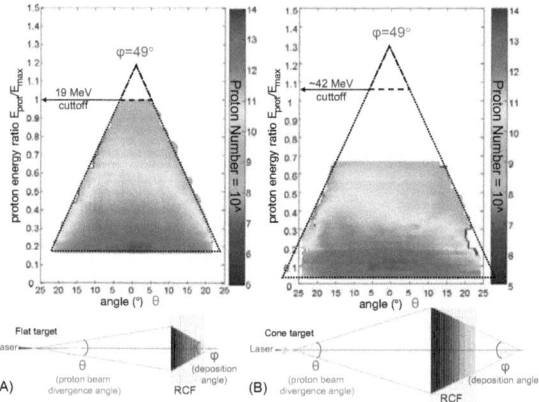

Fig. 4.11: RIS beam reconstructions of (A) a Au 10 μm flat-foil target and (B) a Au 10 μm flat-top cone target. Color indicates the number of protons. The x-axis is the divergence angle, θ, of the beam from the target, and the y-axis is the scaled proton energy $E_{\text{proton}}/E_{\text{max}}$; for (A) E_{max} is 19 MeV and for (B) E_{max} is 40 MeV. Superimposed over each of the reconstructions are similar equilateral triangles with an apex angle of 49°, indicating that the deposition angle φ of the beams is similar even though (B) starts out at a much large initial divergence. At the top of each triangle is a line at the cut-off of (A) and where this cut-off (assuming the typical cut-off of 5° divergence) should be for (B), indicating that the beam was potentially around 42 MeV. Figure taken from ref. [4].

the bar on the right of each figure. Figure 4.11A shows data from a Au flat-foil target and is normalized to an E_{max} of 19 MeV. Figure 4.11B shows data from the Au flat-top cone and is normalized to an E_{max} of 40 MeV, chosen based on simulation data and the fact that, with all things being equal (i.e., laser duration, energy, spot size, beam quality) other than the target shape, as seen previously [11, 72], the TNSA sheath follows a self-similar expansion after the laser is gone, producing proton beams with similar divergences. This is backed up by the majority of RIS data, showing that the proton beam typically becomes undetectable when the beam divergence reaches about 5°-6° as is seen in figure 4.11A. Thus, armed with these two facts an angle to the beam's change in divergence θ (beam size) can be fit, as the beam deposits energy in the RCF stack. This has been termed the deposition angle φ and from the collection of RIS data it is known to be similar for TNSA produced beams, though each beam may have a different initial divergence angle θ due to differences in sheath strength. The bottoms of figure 4.11A and B show representations of the beams diverging from their sources and stopping in RCF. The divergence angle θ and the deposition angle φ range between 40°-60° and 37°-

49° respectively. Dotted similar triangles with apex angles of 49°, corresponding to the deposition angle φ of the Au flat-foil target, have been superimposed over the beam contours in A and B to show that the probable cut-off energy (where the beam divergence would be about 5°) is about 42 MeV for the flat-top cone target B, assuming a divergence of 60°, which is slightly greater than the angle the RCF could measure. The deposition angle of 49° is conservative, as fitting smaller angles such as 37° increases the estimate to about 53 MeV.

In summary, the sheath accelerating the protons from a flat-top cone target is generated from a hotter (since $E_{\text{prot, max}} \sim k_B T$) and denser source ($N_{\text{prot}} \sim n_e$). It is able to impart more energy to more ions over a larger area, leading to the larger observed divergence angle (θ) of about 60° full angle, and more energetic protons, probably ≈ 44 MeV, than compared to a flat-foil. The conclusions are in agreement with 2D-PIC simulations, discussed in the literature [4].

4.3.2 Beam smoothing due to target thickness

The last section presented an energy and conversion efficiency enhancement by mounting a flat foil onto a cone, i.e., by changing the interaction geometry at the front side. Going further to the proton-emitting rear side, the question arises how the electron transport during the target affects the TNSA. Since the relativistic electron transport in shortpulse laser-matter interaction is still not very well understood, only one very obvious topic will be investigated, namely the thickness of the target. Other target parameters, e.g. the use of conducting foils or insulators, have already been investigated by refs. [54, 76].

The experiments were performed at the TRIDENT laser facility. The resulting proton beam profiles with an elliptical laser focus are shown in fig. 4.12. The laser beam profile is shown in fig. 4.12a with $(108.5 \times 28)\,\mu\text{m}$ axes length and was the same for all shots, as well as the laser energy and pulse duration. For comparison only RCF images of 5 MeV protons are shown. Subfigures b to d show the results with the target thickness increasing from 13 μm (b) via 20 μm (c) to 50 μm (d). The colormap was optimized for maximum contrast of each film individually. For an enhanced visibility of the grooves imaged in the RCF semi-transparent lines have been overlapped in b)-d). In d) the target was rotated.

Whereas for thin targets (13 μm) the elliptical laser beam profile is the most dominant part of the proton beam (see fig. 4.12b), for thicker targets the elliptical part becomes less pronounced and the envelope becomes rounder. The source sizes of all

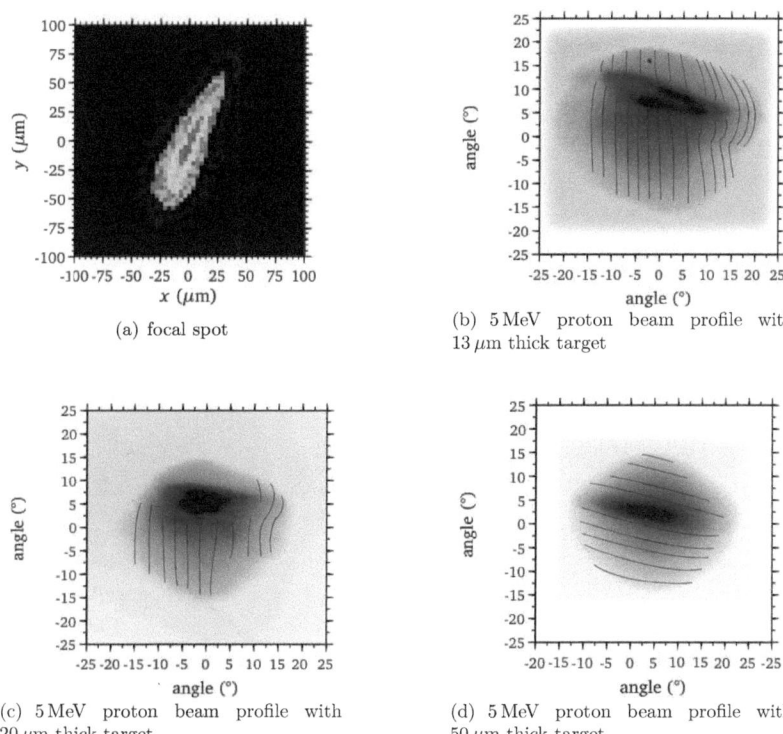

Fig. 4.12: Influence of target thickness on proton beam profile. (a) The laser was defocused to an ellipse with eccentricity 4 and an intensity of $I_L = 2.5 \times 10^{18}$ W/cm^2. (b) For a 13 μm thick target the ellipse in the proton beam profile is clearly visible, the image shows 5 MeV protons. The beam profile of the 5 MeV protons becomes more symmetric with increasing target thickness from 20 μm (c) to 50 μm (d), and the elliptical part moves to the center. In all three cases the laser focus was kept unchanged. For an enhanced visibility of the grooves imaged in the RCF semi-transparent lines have been overlapped in b)-d). In d) the target was rotated.

5 MeV protons are (130 ± 10) μm, (110 ± 30) μm and (90 ± 20) μm for the 13, 20 and 50 μm targets, respectively. It is interesting to note that the 5 MeV proton's source size *decreases* with target thickness. The divergent electron transport [182, 224] should lead to an increasing source size with target thickness. The maximum energy was approximately 7 MeV in all three cases, but with weaker and smaller beam imprints of the maximum energy protons with target thickness. The maximum energy cannot be determined accurately, since the RCF stack's energy resolution of ≈ 1 MeV is weak and therefore is rather subject to error. It is known from the

experiments described above, that protons have a decreasing source size with increasing energy. In the case here the beam spot of protons has slightly decreased with target thickness. Since only 5 MeV protons are considered, independent of the maximum energy, it is assumed that the data for thicker targets represent protons closer to the maximum energy, therefore their source size must decrease with target thickness. The RCF stack's energy resolution was not sufficient to fit a Gaussian to the measured data, therefore a FWHM source size as in the cases above could not be deduced. It can be estimated, however, since the source size of the 5 MeV protons is known and the maximum energy is between 6 and 8 MeV. Therefore 5 MeV energy represents protons between 62.5 % and 83 % of the maximum. Knowing their source size allows to calculate the FWHM source size. It is $\approx (140 \pm 30)\,\mu$m for the 50 μm target and $\approx (200 \pm 45)\,\mu$m the 13 μm thin foil. Therefore, it is concluded, that the whole source size decreased with increasing target thickness, although the electrons have divergently propagated through the target, which resulted in a larger area of electron presence. Since there is no difference on the front side for all three cases, i.e., laser parameters as well as the target's front side were not changed, the same amount of energy was transported from the laser to the rear side. However, for thicker targets the volume occupied by the electrons is larger, hence the electric field strength became less with increasing target thickness. Therefore 5 MeV energy was closer to the maximum with increasing thickness, which explains the decreasing source size with increasing thickness.

In summary, the increasing target thickness results in a smoother proton beam profile with less impact of the laser beam profile, i.e., the elliptical part of the beam becomes less pronounced, moves more to the center of the beam profile and the background becomes more symmetric with target thickness. The influence of the target thickness on the accelerated proton beam profile can be investigated with the SABRINA code, results will be presented in section 5.1. The experimental data, together with the laser beam impact and theoretical investigations with SABRINA have been published in ref. [2].

4.3.3 Large scale curvature

In the previous sections experiments were presented, where the front side of the foil was shaped like a cone, resulting in an enhanced energy conversion efficiency. The divergence of the proton beam was unchanged, with still large opening angles for low energies. This is detrimental for further applications. By shaping the laser focal spot, the divergence of the full beam could not be significantly reduced as well, although the intensity distribution in the beam could be modified. By using thick targets the laser focal spot impact can be reduced. However, just after the discovery of TNSA it was recognized, that on short spatial scales the proton beam can be focused by curving the target rear side [52]. After first tests of the focusability [284], half-sphere-shaped targets were used to ballistically focus the protons and to heat a second target by the proton energy deposition [29]. However, the focal length of these targets is very short, on the order of the sphere's diameter, which limits their use for applications. Another possibility for focusing the protons is the ultrafast laser-driven microlens [81] that uses a second high-intensity laser beam to create a hot plasma expansion towards the axis of symmetry inside a tiny cylinder. The significantly time-dependent electric field of this plasma is used to focus protons traversing the cylinder. The experimental scheme suffers from a complicated geometry with two synchronized high-intensity laser beams that need to be carefully aligned and temporally adjusted. Kar et al. [82] recently succeeded in partly collimating the proton beam by combining a microlens device with a flat target foil into a single piece, at the expense of a complex target assembly.

It is known [29, 284], that strongly curved targets can be used to focus TNSA-protons, whereas flat targets emit the protons with large opening angles. Additionally, the energy-dependent source size follows a Gaussian (see fig. 4.4), therefore it is intuitive that for a beam collimation the curvature of the target should increase like a Gaussian, too. This assumption is strongly supported by the plot of the half opening angle versus source size,

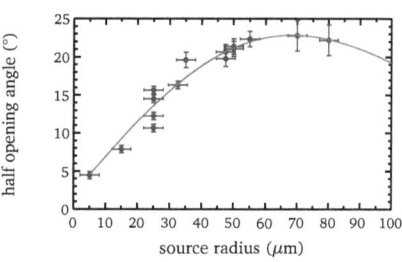

Fig. 4.13: half opening angle versus source radius at Z-Petawatt.

shown in figure 4.13 $(-\circ-)$. The data were obtained at the Z-Petawatt laser, with $E_L = 35\,\mathrm{J}$, $I = 2 \times 10^{19}\,\mathrm{W/cm^2}$ and a $25\,\mu\mathrm{m}$ thick, microstructured gold target. The RCF stack configuration is described in section 4.1.1. The maximum proton energy of this shot was $(20.3 \pm 1)\,\mathrm{MeV}$. Protons with the highest energy had a source size of $5\,\mu\mathrm{m}$ radius and were emitted with an opening angle of $4.5°$. Protons with larger source sizes (and lower energy) were emitted with increasing opening angles up to about $23°$. From fig. 4.4 and its analysis it is known, that the electron sheath can be well approximated by a Gaussian distribution. Following the arguments given in ref. [76], the emission angle of the protons can be obtained as follows: the transverse part of the electric field is proportional to the total field times the transverse gradient of the initial hot electron density: $E_\perp \propto E \nabla_\perp n_e$. The emission angle is thus $\vartheta_\perp(r) \approx p_\perp/p_z \approx \int E_\perp \mathrm{d}t / \int E_z \mathrm{d}t$. The longitudinal field E_z is proportional to $\sqrt{n_e}$ (eq. (2.38)). Hence, the opening angle of the protons (neglecting the source size for large distances) can be determined by the derivative of a transversally Gaussian hot electron distribution $n_e(r) = n_0 \exp(-r^2/2\sigma^2)$, divided by the square-root of it, as

$$\vartheta(r) \propto \frac{r}{\sigma^2} \left(e^{-r^2/2\sigma^2} \right)^{1/2}. \qquad (4.5)$$

A fit of this equation to the data is shown as the red line in fig. 4.13. The FWHM of the fit is $\sqrt{8 \log 2}\,\sigma = 105\,\mu$m. This value is within the error bars of the FWHM determined by the fit to the energy-resolved source size of $92.8\,\mu$m. Both data sets strongly indicate that the hot electron sheath has indeed a Gaussian transverse profile, and the observed angle of beam spread is a result of the Gaussian distribution.

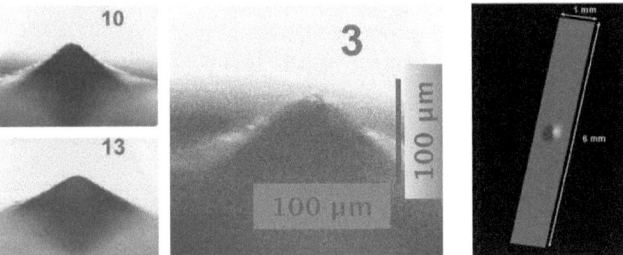

Fig. 4.14: Gaussian shaped target foils. The $10\,\mu$m thin Cu foil has a Gaussian-shaped impression of about $100\,\mu$m width and height. The photographs were taken by a light microscope. The drawing on the right side shows the configuration for the experiment.

These findings lead to the conclusion that Gaussian-shaped target foils could be used to collimate TNSA-protons. The FWHM of the Gaussian foils should be on the order of $100\,\mu$m. Following these suggestions, the foils were made by NanoLabz [282] by order of Sandia National Laboratories for the experimental campaign at Z-Petawatt in december 2008. The targets were made of $10\,\mu$m thick Cu foil, with a Gaussian shaped depression as illustrated in figure 4.14 (right image). Various depressions with varying depths from $0\,\mu$m (flat foil) to about $100\,\mu$m (left images) were made. Unfortunately, due to technical difficulties during the manufacture, the targets did not arrive in time and could therefore not be used in the experiment. Our group decided that the expensive targets should be used in a separate beam time dedicated especially to this experiment at the end of 2008 or later. Therefore, only test-experiments will be presented in this section.

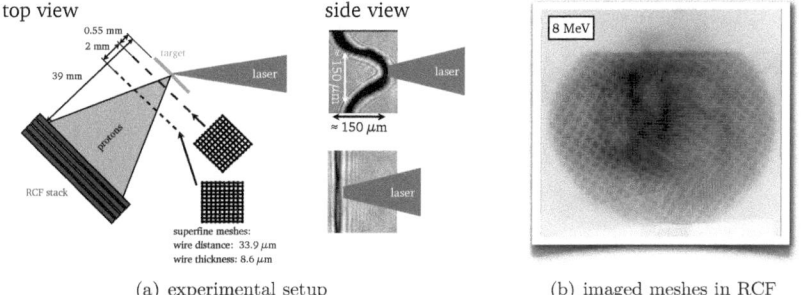

(a) experimental setup (b) imaged meshes in RCF

Fig. 4.15: Point projection configuration. a) The accelerated protons pass two meshes with different orientation and at different distances, that are imaged in the RCF stack, allowing to determine to origin of the point projection. In one experiment, that target was a flat foil, in another one the foil was bent over a wire. The resulting curved foil had about 150 μm height and width, respectively. b) RCF image of 8 MeV protons. The imprints of the two meshes are clearly visible.

Despite a possible collimation of the protons, a focusing effect of the Gaussian shaped foils cannot be detected by the beam size itself, since the focus would be very close to the source. Behind the focus, the beam is again divergent, resulting in a beam spot in the RCF stack that might be indistinguishable from an unfocused beam. Therefore, two fine meshes with 750 lines per inch (lpi) were placed close to the foil, as shown in figure 4.15a. Initially, the foil was flat. The 8.6 μm thin, 33.9 μm spaced wires imprint in the traversing proton beam [25, 209]. The imprint is due to scattering of the protons crossing the material, deflecting them into regions without wires. This results in a shadow of the mesh, imaged in the RCF detector placed at 39 mm distance. The first mesh was placed at 0.55 mm behind the foil, rotated by 45°. The second mesh was placed at 2 mm behind the first one.

The imprinted meshes in the beam profile of 8 MeV protons (E_L = 40 J, I_L > 5×10^{19} W/cm², $E_{\text{prot.,max}} \approx 30$ MeV) are shown in the gray-scale RCF image in fig. 4.15b. The rotated mesh is imprinted with a larger magnification, since it was placed closer to the source. By counting the lines, the beam radii are obtained as $r_1 = 813.6$ μm (24 lines) for the first foil at $d_1 = 550$ μm and $r_2 = 2475.7$ μm (73 lines) for the second one at $d_2 = 2550$ μm, respectively. The averaged periodicity of the lines in the RCF, divided by the mesh periodicity of 33.9 μm, can be used to determine the magnification M_{exp} of the mesh. For the first one, it is $M_{\text{exp}} = 1220/33.9 = 36$ and for the second $M_{\text{exp}} = 412/33.9 = 12.2$.

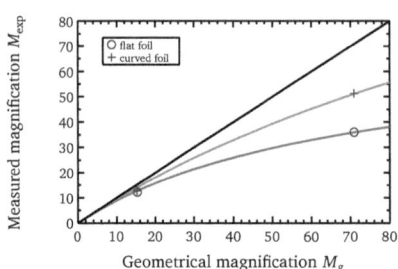

Fig. 4.16: Measured mesh magnification M_{exp} versus geometrical magnification M_g for a flat and a curved foil, respectively.

Now, the method from ref. [74] is used to determine the (virtual) origin of the projection. For a point source, M_{exp} should be the same as the geometrical magnification $M_g = L/d_i$, with the source-to-detector distance L and source-to-object distance d_i. Figure 4.16 shows the measured magnification M_{exp} versus the geometrical magnification M_g. The data points from the two meshes and a flat foil (○) strongly deviate from the geometrical magnification (—). This discrepancy is explained, in that the point source of the projection is not at the target rear surface, but some distance x in front of it [74]. In this case, $M_{\text{exp}} = (L+x)/(d+x) < M_g = L/d$. Therefore,

$$M_{\text{exp}} = M_g (L+x)/(L+M_g x). \tag{4.6}$$

Fitting this function to the measured data (—) results in $x = (550 \pm 50)\,\mu\text{m}$, in good agreement to the data obtained by refs. [9, 61, 74]. However, these results have to be taken with care. More accurate source size data can be obtained by using microstructured targets. Tracing back the lines imprinted in the beam from a microstructured target [9, 61] results in a *virtual source* of about $5\,\mu\text{m}$ diameter, and not a point source. Nevertheless, the projection method can be used to determine whether or not the focusing of curved target foils works.

The geometrical focusing was tested by bending a flat foil over a thin wire, resulting in a quasi-Gaussian curved foil in one direction, that is flat in the other direction. Figure 4.15a shows a shadowgram of the mounted foil before the laser irradiation. The resulting RCF image was similar to the image in fig. 4.15b, however the beam was slightly elliptical with less extension in the vertical direction than in the horizontal one. This is one indication that a collimation of the proton beam by shaping the target works. The second indication is the location of the virtual source. In fig. 4.16 the experimental magnification M_{exp} versus M_g of the curved foil (+) is now closer to the geometrical magnification of a point source, located at the target rear side.

By fitting eq. (4.6) to the data (—), the virtual source position is determined to be at $= (220 \pm 50)\,\mu$m. Hence the position of the point-projection virtual source has moved about $(330 \pm 50)\,\mu$m closer to the target foil. This value is just about twice the height of the foil, indicating that indeed the protons have travelled through a beam waist after leaving the target. This is in close analogy to the results obtained with hemispherical targets [29, 79, 80], where the position of the proton beam waist was found to be very close to twice the sphere's radius [80].

Further investigations should determine the position of the virtual source for all energies measured with the RCF stack. It is assumed, that different energies will seem to originate from different virtual source positions, due to their different divergence (fig. 4.5). Additionally, protons from different source positions will have different opening angles due to the Gaussian-shaped hot electron sheath (fig. 4.13). Hence hemispherical targets like those used in refs. [29, 79, 80] will result in different focal positions for different energies, in analogy to chromatic aberrations of an optical lens. This could be compensated by shaping the target as a Gaussian, that will therefore act as an achromatically corrected proton lens. The experimental realization, unfortunately, has to be left for a future work.

4.4 Beam control with magnetic fields

The previous sections reported about optimization and beam control by modifying either the laser parameters or the target geometry. Changing either one or both of these properties has a direct influence on the acceleration process. This could have the disadvantage, that an improvement of one parameter, e.g., the maximum energy in case of a cone-shaped target front side, leads to a deterioration of another parameter, e.g., the smoothness of the beam profile and the emittance. A crucial parameter, currently hindering the broad application of TNSA-protons is the large opening angle. Gaussian shaped foils could be used to collimate the beam, however the experimental demonstration still has to be realized.

It is well-known that magnetic fields are ideally suited for guiding an ion beam, due to the Lorentz-force $\boldsymbol{F} = q\,\boldsymbol{v} \times \boldsymbol{B}$. For relativistic velocities $\boldsymbol{v}$, magnetic fields $\boldsymbol{B}$ are more efficient to modify the particle trajectory than electric fields $\boldsymbol{E}$. Additionally, the work done on a particle by a magnetic field is zero, hence the particle velocity is not changed in a magnetic field. At first glance it is obvious that magnetic fields could be used to modify the trajectory of TNSA-protons, to control the opening angle of the ions. On second thoughts the use of magnetic fields might cause problems, since the protons are not emitted from the source as a singly-charged ion beam, but in form of a quasi-neutral plasma (sec. 2.4) with electrons accompanying the protons. The removal of the electrons by the magnetic field could lead to strong space-charge forces, leading to a Coulomb explosion of the remaining protons. This would ultimately destroy the beam emittance, as well as it would increase and not decrease the opening angle. Therefore it was tested first, if the electrons can be safely removed from the beam without destroying it. For this experimental test, a magnetic dipole was used to remove the electrons from the beam. The experiment is similar to the experiment reported in refs. [72, 83]. However, in the experiment described here the magnet was placed much closer to the source. The experimental set-up and the results are described in the next section. In a second step, described in section 4.4.2, the beam was injected into a magnetic quadrupole beam line, demonstrating the *world-wide first beam transport and focusing of laser-accelerated protons*.

4.4.1 Magnetic electron removal

The experimental set-up is shown in fig. 4.17. The left image shows a schematic of the experiment, the right image shows a photograph of the set-up in the target chamber. The experiment was performed at the *Z*-Petawatt laser, with similar laser

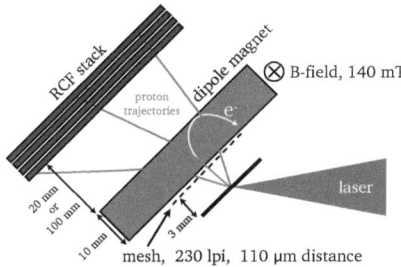
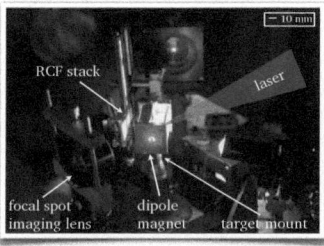

Fig. 4.17: Schematic (left) and photography (right) of the electron-stripping experiment. The laser-accelerated, quasi-neutral electron-proton plasma expands through a wire mesh into a dipole magnet. The electrons are kept back by the magnetic field with 140 mT field strength, whereas the protons pass the magnetic field on nearly straight trajectories. An RCF stack, wrapped in Al-foil, is used to detect the protons and the imaged mesh.

parameters as already mentioned above. The target was a 25 μm thick Cu-foil. At (3 ± 0.5) mm distance behind the foil, a wire-mesh with 230 lines per inch (110 μm wire distance) was placed in the proton beam path. The mesh was fixed at the entrance of the dipole magnet, consisting of 47 mm long, 10 mm wide and 12 mm high permanent magnets. The two magnets had a 15 mm gap for the proton beam. The magnetic field strength was measured with a Hall probe to be 240 mT on the surface of the magnets and 140 mT in the center of the gap. The magnetic field vector is pointing downwards in the schematic in fig. 4.17 and from top to bottom in the photography. The RCF stack for the proton detection was placed either at 2 cm behind the magnet or at 10 cm behind the target (7.7 cm behind the dipole).

A charged particle moving in a magnetic field is forced on a trajectory with the gyroradius $r = p/qB$, where p denotes the particle momentum, q its charge and B the magnetic field strength, respectively. For the dipole magnet in the experiment, the gyroradius is $r = (1-3.2)$ m for (1-10) MeV protons. The trajectories in fig. 4.17 are exaggeratedly drawn for clarity.

The electrons are co-moving with the protons, hence their velocities are equal. The velocity of 10 MeV protons is $v = 4.4 \times 10^7$ m/s, this corresponds to a kinetic energy of the co-propagating electrons of $E_{\text{kin}} = 5.5$ keV only. The corresponding electron gyroradius is $r = 1.8$ mm, that is much less than the 10 mm width of the dipole magnet. The field of 140 mT is strong enough to remove the co-propagating electrons from the quasi-neutral plasma even as close as 3 mm from the target.

The experimental results are shown in fig. 4.18 for 8 MeV protons. Two shots with equal laser and target parameters are compared. In the first shot, the RCF stack

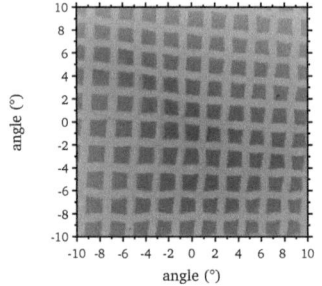
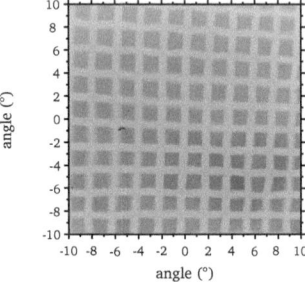

Fig. 4.18: Mesh imprint for 8 MeV protons 2 cm (left) and 10 cm (right) behind the dipole.

was placed 2 cm behind the dipole. The left image of fig. 4.18 shows the mesh imprinted in the beam, for a section with ±10° size. The structure is clearly visible and shows a good contrast. The right image of fig. 4.18 shows the imprint with the RCF placed at 7.7 cm distance from the dipole. Despite the less coloration due to the larger beam expansion, the section with ±10° size is nearly identical to the first shot. In both data sets, the total opening angle of the beam is $\alpha = (22.5 \pm 0.5)°$, that is the same magnitude as the comparison shot without magnets (see fig. 4.3, at $E/E_{max} = 0.4$). A possible Coulomb-explosion of the beam after the electron removal would result in an increased beam spot at the RCF, and therefore a larger opening angle. Since this seems to be not the case, this result is one indication that the removal of the electrons by the magnetic field does not affect the protons.

A better indication of a change in the opening angle is the determination of the virtual source of the point projection as performed in the section before. For the 2 cm distance shot, the virtual source position is determined to be at $z = [-300, +750]\,\mu m$, for the 10 cm RCF distance the virtual source is at $z = (0 \pm 500)\,\mu m$. The errors are very large, due to the error of the mesh position at $z = (3 \pm 0.5)\,mm$. The mesh was fixed on the dipole entrance, the whole device could not be placed more accurately. Compared to the virtual source position of the experiment without any magnet (sec. 4.3.3) at $z = (550 \pm 50)\,\mu m$, both data sets with the dipole in place indicate that the virtual source has slightly moved towards the foil, leading to a slightly larger magnification of the mesh. This would mean, that the opening angle has slightly increased due to the Coulomb repulsion of the de-neutralized beam. However, a clear answer requires additional and more accurate experiments.

Another indication that the proton expansion is not affected by the electron removal is the still excellent contrast of the imprinted mesh. The repulsive Coulomb forces between the protons in the de-neutralized beam lead to a smearing-out of the imprinted wire edges. The image quality is still excellent, leading to the conclusion that the Coulomb repulsion is quite weak and therefore the electron stripping does neither diminish the beam quality nor does it significantly change the opening angle of the beam.

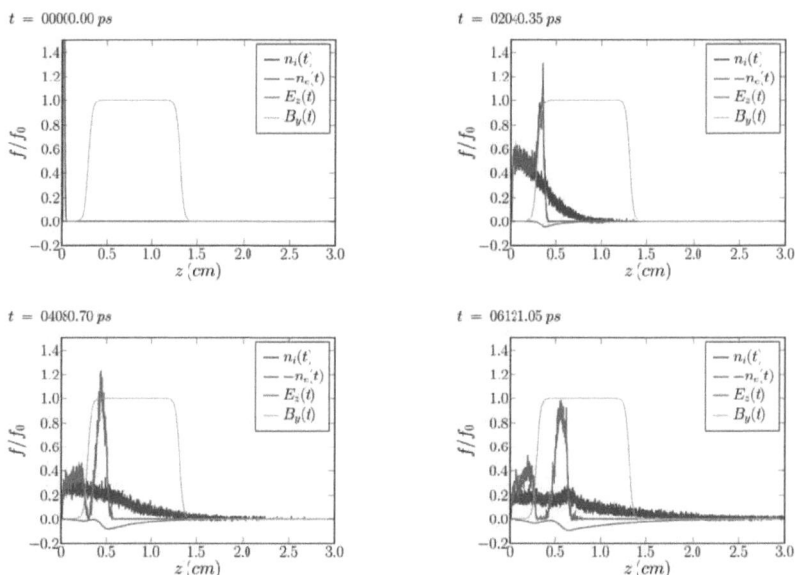

Fig. 4.19: 1D-PIC simulation of a plasma expanding into a magnetic field. Initially, the electron (n_e) and ion (n_i) densities are equal. Later on, the electrons are stopped in the magnetic field B_y. Due to the charge separation an electric field E_z appears.

The analytical estimate from above assumes that the electron-proton plasma is sufficiently thin, i.e., all particles can be considered free. For a denser plasma, the electric fields between the protons and electrons could be too strong and could avoid the stopping of the electrons in the dipole. The scenario was simulated with PSC, since it takes into account the inter-particle forces. The simulation starts with an electron-proton plasma with a Maxwellian velocity distribution, with $N_{(e,i),0} = 5 \times 10^{11}$, a 12 MeV proton temperature and a 6 keV electron temperature, respectively. The two temperatures lead to equal electron and proton velocities. The spatial distribution was set-up with an exponentially decaying profile. An externally applied magnetic field of $B_{y,0} = 140\,\text{mT}$ was implemented in the particle mover in PSC. The magnetic field was placed at $z = 0.8\pm0.5\,\text{mm}$. The field decays as $1/(1+\exp(z))$ at the edges, simulating the fringe fields. The simulation was performed in 1D. The simulation box with a 3 cm length consisted of 3000 cells, with $10\,\mu\text{m}$ per cell. Due to the low density and high temperature, the Debye length is 3.6 cm. The total number of quasi-particles is $n = 6.4 \times 10^5$, with initially 4000 particles per cell. The simulation run for 2.4×10^4 times steps with $\Delta t = 0.37\,\text{ps}$ on one CPU for 48 hours.

The result is shown in fig. 4.19. The upper left image shows the initial distributions of both the electron n_e (—) and ion n_i (—) densities, respectively. The magnetic field B_y (—) stays constant for the whole run. After some time, the protons have started to expand into the magnetic field (upper right image). The electrons, having not enough kinetic energy, are being stopped by the B-field. Even for later times (lower images), most of the protons have already passed the B-field, whereas the electrons are being accumulated at the entrance of the magnetic field. Due to the charge separation, an electric field E_z (—) with $E_{y,\text{max}} = 1.4 \times 10^7\,\text{V/m}$ appears.

4.4.2 Transport and focusing with quadrupole magnets

In this section a straightforward approach is presented that uses an ion optical system consisting of novel Permanent Magnet mini Quadrupoles (PMQ) with strong field gradients of up to $500\,\text{T/m}$, originally developed for laser-accelerated electrons [285]. A set of two mini quadrupole lenses demonstrates transport and focusing of laser-accelerated protons in a very reproducible and predictable manner. This approach uses permanent magnets that do not need to be replaced, hence allowing the application in upcoming high-energy, high-repetition rate lasers, e.g. [286]. Moreover, this approach decouples the acceleration process from the beam transport. This separation allows for independent optimization of the proton beam generation and of the focusing mechanism. The experiment and results are published in ref. [1].

An initial experiment was carried out at TRIDENT, and obtained a line-focus, whereas the demonstration experiments with a point focus were carried out at the 100 TW section of Z-Petawatt. A $25\,\mu\text{m}$ thin Cu-foil was used as target, being hit by the p-polarized laser at an angle of $45°$. The RCF stacks in the experiment consisted of eight layers of type HD-810 and nine layers of MD-V2-55. A scheme of the experimental configuration is shown in figure 4.20. One RCF stack was placed at $(40\pm1)\,\text{mm}$ behind the target, detecting the divergent proton beam. The aperture

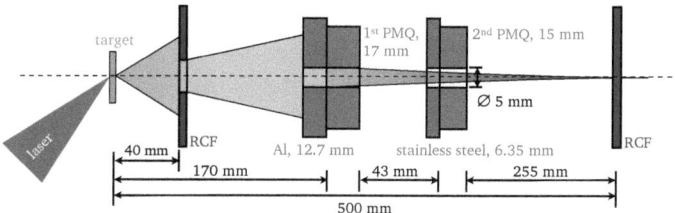

Fig. 4.20: Scheme of the experimental configuration to transport and focus 14 MeV protons. Protons from the rear-side of the 25 μm Cu-foil propagate into the RCF stack with a 5 mm diameter central aperture. This stack detects the initially diverging beam. The protons propagating through the aperture enter two permanent magnet mini-quadrupole devices (PMQ) that transport and focus the beam 500 mm behind the target. A second RCF stack in the focal plane records the energy-resolved intensity distribution of the protons.

was 5 mm in the center where the protons enter the PMQ beam transport section with a 5 mm aperture in the center as well. Beam blocks consisting of 12.7 mm aluminum or 6.35 mm stainless steel protected the PMQs from debris and unwanted irradiation. The magnetic fields were calculated by S. Becker using a Maxwell-compliant solver for their specific design [287]. These fields were used to determine the positions of the PMQs and the spectrometer with a tracking algorithm [288].

The goal was to focus 14 MeV protons, since this energy is in the central region of the proton energy spectrum usually produced at TRIDENT and Z-Petawatt. The first PMQ with a 17 mm length was placed at a distance of 170 mm behind the target, the second quadrupole with a 15 mm length was placed 43 mm behind the first one. The focal spot was expected 500 mm behind the target, where another RCF stack was placed. The maximum proton energy was well above 22 MeV, which was the upper detection limit of the RCF stacks used. The total number of particles and their energy spectrum were obtained from the first RCF stack by interpolating over the aperture in the center. The resulting particle number spectrum dN/dE per unit energy follows the shape from eq. (3.15) with parameters $N_0 = 4.9 \times 10^{12}$ and $k_B T = 1.24$ MeV.

A typical beam profile of (14 ± 1) MeV protons is shown in Fig. 4.22(a). The white spot in the center is due to a hole allowing for the propagation of the protons through the PMQs. The beam profile shows intensity modulations that originate from micro-corrugations of the not-so-perfectly flat target rear surface. The beam has a diameter of (29.5 ± 2) mm that corresponds to a $(20° \pm 1.5°)$ half opening angle. A summation of the total signal in Fig. 4.22(a) leads to a total number of 1.3×10^{10}

protons with (14 ± 1) MeV. About 7.5×10^8 protons entered the PMQs. This number corresponds to 7.5 % of the beam injected into the PMQs. The integration over the spectrum yields a conversion efficiency of 1 % of the laser energy into protons with energies above 4 MeV, in agreement with Ref. [15].

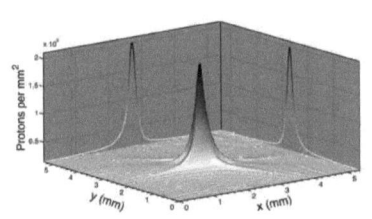

Fig. 4.21: Surface plot of the focused protons from fig. 4.22(b).

The focusing effect of (14 ± 1) MeV protons 500 mm behind the target is shown in Figs. 4.22(b) and 4.21. The color-map in (b) was optimized to show the weak background signal, therefore the film appears to have a signal over the whole area. By integrating over the peak, a total number of 8.4×10^5 protons is obtained. Hence the transmission through the magnets was $8.4\times 10^5/7.5\times 10^8 = 0.1$ %. This was expected, since the first PMQ focused the beam in one plane and defocused the protons in the perpendicular one. The second PMQ's aperture then cut most of the beam.

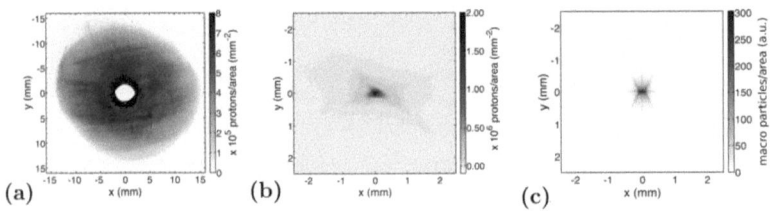

Fig. 4.22: (a) Beam profile of (14 ± 1) MeV protons at 40 mm behind the target. (b) Beam spot in the expected focus 500 mm behind the target. (c) The simulation with a tracking algorithm with (14 ± 1) MeV, neglecting emittance, shows good agreement with the experiment.

Although the PMQs were not especially designed for this beam, a small focal spot was obtained. The spot size was by far not limited by the emittance, which is on the order of $[10^{-2}-10^{-3}]\,\pi\,\text{mm}\,\text{mrad}$ [72]. Simulation results by S. Becker from LMU München show good agreement with the experiment (Fig. 4.22(c)). The RCF was simulated using an ion beam with a Gaussian initial energy distribution of $E=14$ MeV and standard deviation $\sigma=1$ MeV. The PMQ's aperture encircles the solid angle of the ion beam by orders of magnitude. It justifies the assumption of a uniform initial particle distribution within a much smaller solid angle in order

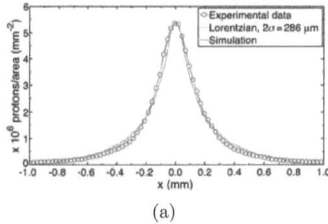
(a)

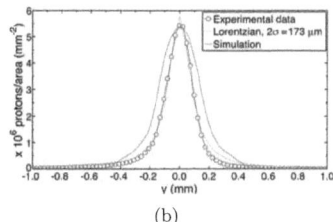
(b)

Fig. 4.23: (a) Horizontal intensity profile through the beam spot in Fig. 4.22b. (∘) shows the experimental data, (—) represents a line-out through the simulated beam profile. Both curves agree well with a Lorentzian fit $f(x) = \sigma/(x^2 + \sigma^2)$ (—)) with a FWHM of $2\sigma = 286\,\mu$m. (b) The vertical profile fits a Lorentzian (—) with a FWHM of only $2\sigma = 173\,\mu$m.

to achieve the best possible statistics for the simulation. The number of macro particles was 10^6. Interactions (i.e. space charge) were neglected. The horizontal and vertical line-outs of both experiment and simulation (Fig. 4.24) can be well described by a Lorentzian $f(x) = \sigma/(x^2 + \sigma^2)$ with FWHM $2\sigma = 286\,\mu$m horizontally and $2\sigma = 173\,\mu$m vertically, which corresponds to a decrease of the proton beam compared to an unfocused beam of approximately 10^3 times. An estimate based on the simulations for an optimized set-up, suggests a decrease in the focal spot by an additional factor of 5 in both planes, leading to a demagnification by a factor of $5 \times 5 \times 10^3 = 2.5 \times 10^4$. Such a small focus is still below the space-charge limit, which can be estimated by the generalized perveance, entrance radius and focal spot size [280]. Assuming a pulse duration of 0.7 ns from the drift difference of $(14 \pm 1)\,\text{MeV}$ protons, the space-charge limit for an optimized focus with 5 mm aperture PMQs corresponds to $\approx 10^9$ protons.

The chromatic properties of PMQs yield a focal spot size dependent on the proton energy. Fig. 4.24(a) shows the ion energy spectrum integrated over an area of $200\,\mu$m in diameter. The open circles represent the measured data using the PMQ doublet and the red line displays the calculated proton spectrum for comparison, using the first RCF stack, under the assumption of an undisturbed propagation of the beam to the same distance. The energy-dependent flux increase due to the focusing is shown in Fig. 4.24(b). The peak around $(14 \pm 1)\,\text{MeV}$ protons can clearly be seen. For this specific PMQ-configuration, the flux increase for $(14 \pm 1)\,\text{MeV}$ protons peaked at about a factor of 75. This allows the system to be used as a spatial filter in order to monochromatize the ion energy spectrum. For an optimization and increased coupling efficiency into the ion optics section, the magnets can be placed closer to the source in combination with an increased aperture of the second PMQ. The latter becomes necessary due to defocusing of the first PMQ in one plane, as well as the space-charge limitation mentioned above that decreases with increasing entrance radius.

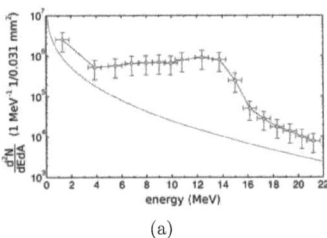
(a)

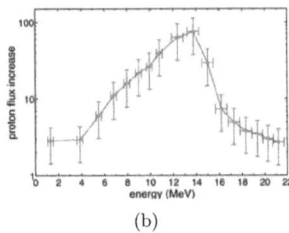
(b)

Fig. 4.24: (a) 500 mm behind the target, in a small area with 200 μm diameter, the proton spectrum without the PMQs exhibits the usual spectrum (—). In contrast, the spectrum with PMQs shows a strong signal enhancement, that peaks at the precise design value of 14 MeV. (b) The proton flux peaks at 14 MeV with nearly 75 times more protons per area compared to the case without PMQs.

Chapter 5

3D proton expansion model

The previous chapter reported the experimental results obtained for this thesis. It was shown, that the profile of the laser focal spot can be used to partially shape the proton beam. This laser beam imprinting effect can be compensated by increasing the thick targets. With thicker targets, the proton beam becomes smoother and the laser beam imprint becomes weaker. More general, it was demonstrated in the last chapter that the properties of TNSA-protons are very similar, independent of the laser system. The beam parameters spectrum, opening angle and source size always show a similar shape. Current one-dimensional plasma expansion models show the affinity to a quasi-neutral, self-similar plasma expansion. However, up to date there is no published three-dimensional model that can be used to reproduce all parameters of TNSA proton beams. Instead, one still has to rely on time and computing-power consuming simulations. The plasma expansion models (sec. 2.4) can only be used to determine one-dimensional parameters as spectrum and maximum energy, respectively.

The time history of the ion acceleration process can be roughly divided in three stages [86]: (i) before substantial ion motion sets in; (ii) the expansion stage (free drift) and (iii) an intermediate stage where the ions get most of their energy. The duration δt of the first stage, where the atoms have already been ionized but have not started moving, can be estimated by the time the ions need to move a distance of the initial Debye length λ_D in the field E_0 (eq. (2.43)): $\delta t = \sqrt{m_p \lambda_D/(eE_0)}$. For the parameters of the standard case from chapter 2 ($n_{e,0} = 1.4 \times 10^{20}\,\text{cm}^{-3}$, $k_B T = 1\,\text{MeV}$) this time is $\delta t = 54\,\text{fs}$. This time is negligible compared to the total acceleration time of about $1\,\text{ps}$.

In the following, first a model of the second stage (free expansion) after the acceleration will be developed. The model is based on a quasi-neutral plasma expansion. It can be used to determine the opening angle and beam shape of a single proton energy. Furthermore, it has been used to explain the laser focal spot impact as well as the influence of the target thickness on TNSA-protons [2].

For the intermediate stage – the actual acceleration process – an empirical model is presented in sec. 5.2 that is able to reproduce all parameters of TNSA-protons. Furthermore, the model can be used to explain the imaging of target surface grooves into the RCF stack and to reconstruct the hot electron sheath in 3D.

5.1 Sheath-Accelerated Beam Ray-tracing for IoN Analysis code - SABRINA

A full 3D-model of the electron generation and transport in the experiments is beyond the capacity of current computer codes and needs a more simplified approach. The model described in this section neglects the actual acceleration process. The model describes the expansion stage, i.e., after the acceleration stage. The model will be used to explain the experiments on the laser beam profile impression and the beam smoothing by the target thickness. It takes into account the electron transport through the target and is used to describe the shape of the proton beam in the RCF stack, depending on the shape of the laser beam and the target thickness. The numerical implementation has been labeled *Sheath-Accelerated Beam Ray-tracing for IoN Analysis code, SABRINA*.

The physical picture is as follows: The laser interacts with the electrons of the pre-plasma formed by the pre-pulse and transfers a large fraction of its energy to hot electrons [167]. The hot electrons are approximated by a Boltzmann distribution with a temperature given by eq. (2.28). The hot electron temperature depends on the laser intensity only. Since for the elongated astigmatic foci there is no simple way to obtain the intensity by taking just the FWHM, the intensity of the laser pulse was calculated by counting all pixels above 50 % of the maximum in the recorded focal spot images (see fig. 4.12a for an example). This value is used for the area of the focus. The intensity is then just the laser energy divided by the pulse duration and this area. It was assumed that the spatial distribution of the hot electrons closely follows the laser beam profile, with an average electron energy that is calculated with eq. (2.28). The huge amount of MeV-electrons results in a mega-ampere electron current that is transported through the target. Integrated 3D hybrid-PIC simulations of laser interaction and electron transport with thin foils (10 µm) show that the electrons propagate nearly ballistically through the target [289], whereas 3D-simulations by hybrid-PIC without laser-interaction and thicker foils [222] indicate a collimated transport guided by magnetic fields and possible filamentation of the electron current into beamlets. However, most of the experimental results (see refs. [182,224] for a short summary) show a divergent electron transport. Therefore, and for simplicity, a possible collimation by a magnetic field is neglected. Nevertheless, due to the solid-state density of the cold target there will be collisions by the electrons with the background material that tend to broaden the electron current distribution. This contribution of multiple small-angle scattering leads to an angular broadening of the electrons. In this model Moliéres theory of multiple scattering by Bethe [225] is used to calculate the broadening of the electrons. The angular broadening is given by eq. (2.30).

After passing the target the electrons form the dense sheath at the rear side that accelerates the protons. In the framework of an electrostatic, quasi-neutral ideal two-fluid model [244] the electric field driving the expansion only depends on the gradient of the electron pressure (eq. (2.49)):

$$\boldsymbol{E}_{es} = \frac{-\nabla p_e}{en_{\text{hot}}} = -\frac{\nabla (n_{\text{hot}} k_B T_{\text{hot}})}{en_{\text{hot}}}. \tag{5.1}$$

The angular direction of the protons therefore depends on the gradient of the electron sheath:

$$\alpha_{\text{prot}} \propto \frac{\nabla n_{\text{hot}}}{n_{\text{hot}}}. \tag{5.2}$$

The algorithm takes the laser beam profile as input. The broadening due to small-angle scattering is represented by convolving the electron distribution (i.e., the laser focus image) with a Gaussian angular distribution with a FWHM angle from Moliéres theory at an energy that corresponds to $k_B T_{\text{hot}}$. The result represents the electron sheath at the rear side; this is then divided into a grid. The normal direction is then calculated for each grid element. In this approximation of a quasi-neutral plasma the longitudinal electric field is proportional to the height of the sheath. Therefore the higher energy protons originate from the outer region (in longitudinal direction) of the sheath.

Similar to the model by Ruhl et al. [87], the effect of the grooves at the rear side was included as a sinusoidal perturbation $\delta\alpha$ in the angle of expansion, with a periodicity

λ of 5 µm or 10 µm and an amplitude A fit to the data (the amplitude just controls the visibility of the lines), as

$$\delta\alpha = A \sin\left(\frac{2\pi x}{\lambda}\right). \tag{5.3}$$

Depending on the initial line orientation at the target the perturbation was taken either in x- or in y-direction.

The height of the sheath cannot be calculated by taking the laser beam profile measurements. Since the protons expand in direction of the sheath normal, the measured angle of beam spread is used to adjust the height of the sheath until the simulated beam fits the measured one. The model therefore allows for a reconstruction of the shape of the plasma sheath at the end of the acceleration, that determines the proton's angular expansion.

5.1.1 Application: Electron transport in solids

Fig. 5.1 shows the application of the model to an experiment with a 50 µm Au-foil that had 5 µm grooves at the rear side. Fig. 5.1a shows the measured laser focus with an intensity of 7.3×10^{18} W/cm^2. The ponderomotive potential is 0.75 MeV, that results in a hot-electron temperature of 1 MeV. The full angular broadening of the hot electrons with a kinetic energy of 1 MeV then is 68° FWHM. It has been assumed that only the upper 70 % of the sheath contribute to the 5 MeV protons that were measured in the experiments. This value is motivated be the argument that protons above that 5 MeV are between 62.5 % and 83 % of the maximum energy, as determined in subsection 4.3.2. It is worth noting that below 60 % of the sheath, it has – due to the Gaussian shape – a reflection point that leads to rings in the image as pointed out by Brambrink et al. [290]. Those rings were not seen in our experimental data.

The resulting distribution at the rear side is shown in Fig. 5.1b. A best fit with SABRINA to the experiment is obtained by adjusting the height of this sheath to 15 µm, see Figs. 5.1c and 5.1d for a comparison of 5 MeV protons. The height of 15 µm is comparable to values already published in literature [26] and leads to an acceleration time on the order of a ps. In both model and experiment the source size can be deduced to be (95±15) µm. Although the broadening angle calculated by small-angle scattering reproduces this experimental data point, this value does not fit for different energies of the protons. The maximum proton energy was $\approx$ 7 MeV with a source size of 65 µm. As mentioned before, there could have been higher energy protons, above 7 MeV and below the detection threshold of the next RCF, with less source size. The 7 MeV protons' source size implies an electron transport cone angle of 24°, that is close to the value reported in ref. [76] with a similar target but for a lower proton energy. However, a SABRINA simulation with this small angle does not reproduce the measured beam profile. The measured profile instead can be reproduced by taking the broadening of 68° FWHM and only the upper 5% of the resulting rear side sheath. A test with different shots indeed shows that beam profiles for the highest-energy protons can be reproduced by taking only the upper few percent of the simulated sheath profile. The profiles for lower energetic protons then can be simulated by subsequently taking more of the sheath.

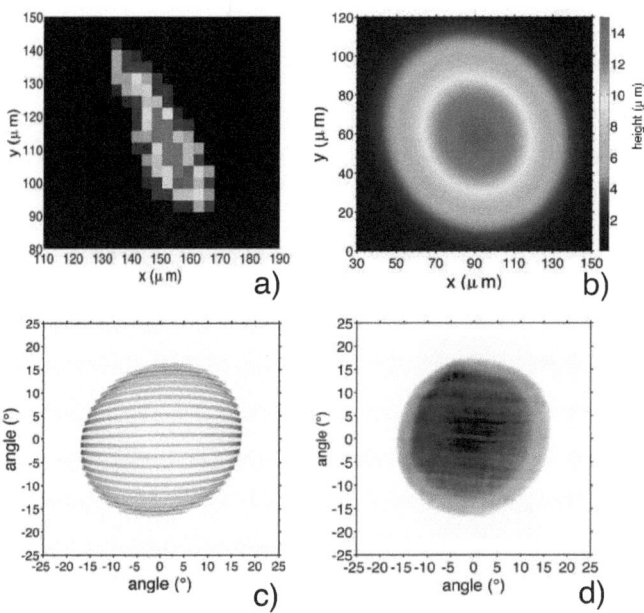

Fig. 5.1: Simulated proton beam profile for a 50 μm thick Au target. The laser focus (a) is taken as an input parameter, the intensity of this shot was 7.3×10^{18} W/cm^2, resulting in hot electrons with a temperature of 1 MeV. The broadening due to small-angle scattering leads to a smooth rear side electron distribution (b), that in turn shows a smooth simulated proton image (c) that closely reproduces the measured 5 MeV proton image (d). The 15 μm height of the sheath was fit so that the divergence of the simulated and measured beam overlaps. The ring visible in (d) is a result of the transmission scan through the two active layers in MD-55.

This leads to the following physical picture: The electron sheath is fully developed before significant expansion of the protons occurs. The protons with the highest energy are accelerated by the strongest part of the electric field that has its maximum in the center. The accelerating electric field amplitude decays like a Gaussian in the transverse direction, therefore lower energy protons originate from larger sources, see Fig. 4.4. The height of the sheath determines the angular expansion.

Next this model is compared to data from experiments with $13\,\mu$m thick Au foils. An experimental result from TRIDENT ($I = 2.5 \times 10^{18}\,\mathrm{W/cm^2}$, $U_\mathrm{pond} = 0.26\,\mathrm{MeV}$, $k_B T_\mathrm{hot} = 0.7\,\mathrm{MeV}$) shall be reproduced. The angular broadening due to multiple small-angle scattering is 42° FWHM. Figure 5.2a shows the measured laser beam profile. The proton beam profile is shown in Fig. 5.2b. Despite the ellipse due to the laser beam profile the lower part of the image shows vertical lines from the grooved target surface. The source size of this part is $130\,\mu$m. In the upper part the former vertical lines were bent to the left, and they have less contrast than the lines in the lower part. The large source size would imply a very large electron transport cone angle of 148° FWHM, that is more than two times the angle by small-angle scattering. Indeed, neither a SABRINA simulation with broadening due to small-angle scattering only (see Fig. 5.2c) nor a simulation with 148° broadening (Fig. 5.2d) coincide with the measured data. The shape of the image produced by the small-angle scattering calculation roughly fits the intense elliptical part of the measured data. The round part is reasonably reproduced by the calculation with the large cone angle.

This could imply that two different proton beams overlap in the temporally integrating detector. If there had been two temporally separated populations however, the straight lines from the large angle part should overlap with the elliptical part and should be visible everywhere in the measured data. A close inspection does not show any lines in the elliptical part, although the RCF was not saturated. Instead, there is a slight bending of the lines, that was even more visible in different shots. Therefore it is concluded that the image was not produced by two different proton beams.

The measurements can be approximated however, by keeping the broadening of 42° by small-angle scattering and additionally by magnifying (interpolating) the laser focus image up to the measured source size. This results in a rear-side sheath that follows the laser beam topology. Fig. 5.2e shows the sheath, the height of $200\,\mu$m was fit to result in a beam divergence that matches the measured data. The resulting proton beam profile is shown in Figure 5.2f and shows a reasonable agreement with the experiment. For this reproduction of the beam profile the source size had to be increased by a factor of three. The most energetic protons of this shot had an energy of 7 MeV with a source size of 80 µm. Again a SABRINA simulation shows that only the case where the upper 5% of the sheath that was fit to the 5 MeV protons is taken shows good agreement. The same is found for thin targets compared to thick ones for the experiments at LULI: The SABRINA-model can be used to reproduce the shape and source size of laser-accelerated protons for thick targets. For thin targets the shape of the sheath can be estimated by the assumption of multiple small-angle scattering, too. However, the size of the resulting sheath must be increased several times (up to a factor 5) to match the measured source sizes.

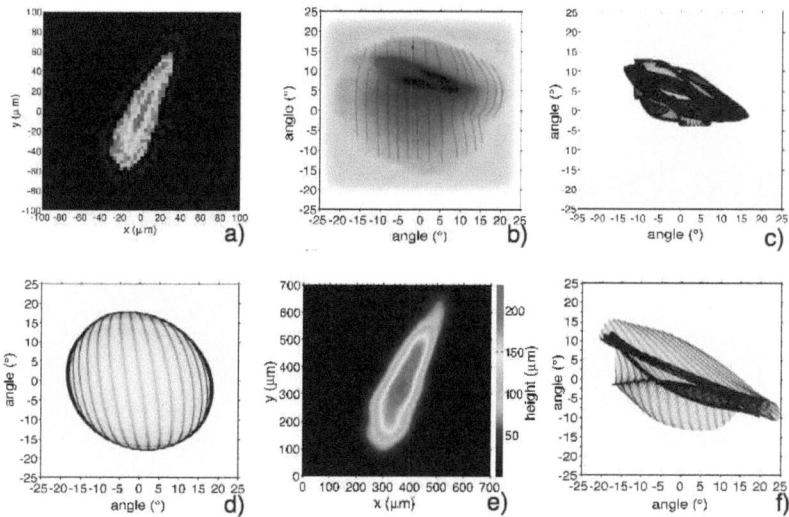

Fig. 5.2: Simulated images for a thin 13 µm Au target. (a) The laser focus was taken as input, (b) shows the measured beam profile for 5 MeV protons. Semi-transparent lines have been overlapped to enhance the visibility of the grooves imaged in the RCF. (c) The simulation with broadening due to small-angle scattering only shows a strongly deformed proton beam profile, (d) whereas a broadening with $\alpha = 148°$ up to the measured source size of 130 µm shows a beam without ellipse. (e) The measurements can be simulated however, by magnifying (interpolating) the laser focus image to the measured source size and by taking a broadening due to small-angle scattering for homogenization. This results in a rear-side sheath that closely follows the laser beam topology. The height of 200 µm was fit to result in a beam divergence that fits to the measured data. (f) The resulting proton beam profile shows a reasonable agreement with the experiment.

Chapter 5. 3D proton expansion model

5.1.2 Discussion

There are several physical mechanisms that can be responsible for the large source sizes that were measured with thin targets. First there could be some transverse spreading of the electrons at the rear side, as reported recently by McKenna *et al.* [63]. The velocity of the spreading observed there is constant with $\approx 3/4\,c$ (where c is the speed of light in vacuum). The sheath for TRIDENT in the section above, is about twice the size of the laser focus. This would correspond to a time duration of the electron spreading of $50\,\mu m/2.25 \times 10^8\,m/s = 222\,fs$ if a uniform electron velocity is assumed. This is a factor of four less than the laser pulse duration, so it seems to have no relation with the laser. Additionally, a constant spread velocity of the sheath would result in an isotropisation of the elongated sheath, since the thin side will – relatively to its initial size – expand much more than the long side, leading to a decrease in aspect ratio. The SABRINA simulation from above, however, shows that the aspect ratio of the sheath closely follows the laser beam. Therefore a transverse spread as reported in ref. [63] seems not to explain the results in this work.

Another possibility could be the existence of two different electron populations like those measured by Gremillet *et al.* [215] and just recently by Santos *et al.* [162]. They report about a fast, strongly collimated (jet-like) hot MeV-electron beam and a second, much broader electron bulk component. Although their measurements were either done with insulating glass targets at high laser energy [215] or aluminum foils at low laser energy [162], two electron populations could exist in the current experiments at high laser energy and gold foils, too. Both electron populations expand fast enough to overlap at the rear side and could contribute to the proton acceleration. However, as stated by ref. [162] the collimated feature contains only very little energy around 1 % of the laser energy, compared to the 35 % energy contained in the bulk part. In addition to that is the bulk's divergence angle of $35° \pm 5°$ determined there too small to explain the findings in our experiments.

The next possibility that could explain the large source sizes for thin targets could be the inhibition of the electron transport [169, 196]. The strong electromagnetic fields generated at the front side inhibit a large amount of electrons from entering the target. It has been determined by numerically modeling the electron transport in Aluminum foils, that for thicknesses below $20\,\mu m$ [169] the electromagnetic fields dominate the electron transport. For larger distances, the transport is dominated by collisions. The foils used in the experiment consisted of Gold, that is much denser. Hence the collision-dominated regime should be closer to the surface. Nevertheless, the computer simulations and experiments [291] have shown that for target thicknesses below 15 μm the targets are much more heated by the slow electrons that are trapped in the electromagnetic fields. This could lead to a larger transverse extension of the hot electron sheath at the rear side. An increased temperature should lead to an increased maximum energy of the accelerated protons. However, this has not been observed in the experiments.

Another reason for the observed source sizes could be recirculation of the hot electrons. Since the sheath field at the rear side is convex (e.g. like a Gaussian), it can act as a weak focusing lens for the reflected electrons. At the front side they can either enter the area of the laser focus again or they will be pushed back into the target by the charge-separation field at the front side. This can lead to an increase of the sheath density and temperature, hence to an increase of the electric field. Electrons with a velocity close to the speed of light can travel back and forth (re-flux) the thin target for about 10 times during the laser pulse duration of $\approx$ 1 ps. The range of 1 MeV electrons in (cold) gold is $\approx$ 150 μm [292], so the electrons keep re-fluxing when the laser is off until their energy is lost. Recirculation effects [293] are much less important for the thick targets where the electrons travel only 3 times through a 50 μm foil while the laser is on. Measurements and modeling of Cu-K_α x-ray emission and electron transport by irradiating copper foils with similar intensities as in this article support this argument [229, 294]. Additionally it is possible that the electrons were injected with a certain angular distribution. The injection depends on the pre-plasma profile as well as on the laser beam topology and therefore is hard to estimate and needs a fully detailed three-dimensional computer simulation.

5.1.3 Conclusion

The Sheath-Accelerated Beam Ray-tracing for IoN Analysis code, SABRINA, has been developed, that takes the laser beam parameters as input and calculates the shape of the proton distribution in the detector. The electron transport was modeled to closely follow the laser beam profile topology and a broadening due to small-angle collisions was assumed. It was shown that broadening due to small-angle collisions is the major effect to describe the source size of protons for thick target foils (50 μm). In contrast to that, thin target foils (13 μm) show much larger sources than expected due to small-angle collisions. The physical reason behind this observation stays unclear and is most likely the result of electron re-fluxing.

These conclusions are even further complicated by a publication early of this year by Green *et al.* [182], showing an increasing electron divergence with laser intensity. The intensities treated in this thesis are all between $I = 10^{18} - 5 \times 10^{19}\,\text{W/cm}^2$. Following their conclusion the fast-electron divergence would be constant, at about $(20-35)°$ FWHM. The collisionless PIC simulations in ref. [182] were interpreted, in that this angle is generated at the laser-interaction zone already, in agreement with the findings of the collisional simulations by Welch *et al.* [289] as well as in the simulations performed in the framework of this thesis, shown in fig. 2.14.

Contradictory to that, an injection angle of $\approx 30°$ is in close agreement with collisional simulations of the electron transport by Honrubia *et al.* [222]. In the reference, an electron distribution with this opening angle was injected into a preformed plasma in front of a solid target, just as depicted in fig. 2.5. Their simulations show first a collimation by the magnetic field (for $z < 10\,\mu\text{m}$) and later on a strong divergence that is attributed to collisions. Hence it can be doubted that the injection angle of the electrons at the front side of the target is identical to the angle measured in ref. [182] by K_α-spectroscopy inside the target, further confusing the interpretation of experimental data. Additionally, computer simulations indicate a collision-dominated transport for thicker targets [169, 222], with the result of a quadratic increase of the electron broadening angle with thickness (see fig. 2.6).

In summary it is concluded, that the shape of the sheath at the rear side of thick targets can be estimated by a simple model of broadening due to multiple small-angle scattering, but it fails for the description of the sheath broadening in thin targets. A clear answer of the physical mechanism behind the large source sizes and sheath profiles requires further full-scale integrated three-dimensional computer simulations containing the laser-plasma interaction as well as the electron transport and further progress in the experimental measurements of the electron transport, which are beyond the possibilities of this work and left for the future.

5.2 3D-model based on flow characteristics

SABRINA can be used to explain the beam profile of TNSA-protons in dependence on the laser beam profile and target thickness. However, only one RCF image and the corresponding sheath height after the acceleration can be determined in one simulation run. A more realistic description of TNSA must include the acceleration process as well as the full beam, including all energies. The framework of a plasma expansion model would include the solution of the fluid equations and the Poisson equation in three spatial dimensions. The numerical solution would most likely be as complex as the PSC, and most likely be not much faster in execution speed. In this section a different approach is presented, that uses much simpler transfer functions for the flow expansion. The model reproduces all features measured with RCF imaging spectroscopy and can be used to fully reconstruct the beam. The model relies on the observations presented in the experimental chapter, as well as on two-dimensional PIC-simulations. The first version of the model has been developed by Hartmut Ruhl [87, 88]. A refined version, as well as the motivation by numerically calculating the expansion of a Gaussian, quasi-neutral plasma has been worked out in this thesis by the author in close collaboration with Hartmut Ruhl.

5.2.1 Essential physics of flow expansion

In this section an empirical method to characterize the expansion of laser-accelerated proton flows is developed. The procedure follows the description in ref. [12], here in more detail. It is possible to understand the essential physics of flow expansion of a plasma plume with an analytical solution by Dorozhkina & Semenov [295, 296]. The evolution of electrons and ions during plasma expansion is described by the Vlasov kinetic equations:

$$\left[\frac{\partial}{\partial t} + \sum_{k=1}^{3}\left(v_k \frac{\partial}{\partial x_k} + \frac{e}{m_e}\frac{\partial}{\partial x_k}\Phi\frac{\partial}{\partial v_k}\right)\right] f_e = 0, \quad (5.4)$$

$$\left[\frac{\partial}{\partial t} + \sum_{k=1}^{3}\left(v_k \frac{\partial}{\partial x_k} - \frac{Ze}{m_i}\frac{\partial}{\partial x_k}\Phi\frac{\partial}{\partial v_k}\right)\right] f_i = 0, \quad (5.5)$$

where m_e is the electron mass, Z and m_i are the ion charge and mass, respectively. $\Phi(\boldsymbol{x},t)$ is the electric potential arising as a result of of the charge separation during the expansion of the plasma, that is consistent with the Poisson equation:

$$\sum_{k=1}^{3} \frac{\partial^2 \Phi}{\partial x_k^2} = 4\pi e(n_e - Zn_i), \qquad (5.6)$$

$$n_{e,i} = \int f_{e,i}(\boldsymbol{v}, \boldsymbol{x}, t) \, \mathrm{d}^3 v. \qquad (5.7)$$

$n_{e,i}$ is the electron or ion density. If quasi-neutrality is assumed, $Zn_i = n_e$, the excitation of plasma waves is prevented. Therefore fast changes in time, like plasma oscillations, will be neglected ∂_t is set to zero. The slow motion in time is still covered by the convective derivative $\boldsymbol{v}\partial_x$.

The electric field can be expressed by equation (5.6), by inserting the distribution functions and by calculating the first moment ($\int v_k \mathrm{d}^3 v$):

$$e\frac{\partial \Phi}{\partial x_k} = eE_k = \frac{m_e m_i}{m_i + Zm_e} \sum_{k=1}^{3} \frac{\frac{\partial}{\partial x_k} \int v_k v_j (Zf_i - f_e) \, \mathrm{d}^3 v}{\int f_e \, \mathrm{d}^3 v}. \qquad (5.8)$$

The strong assumption of quasi-neutrality is usually good for expanding flows, and it is required to obtain simple closed analytical solutions. Murakami and Basko [239] found a self-similar solution for non-neutral expansion, but it is analytically solvable only for special geometries. Quasi-neutrality also forces the mean flow velocities $u_{e,k}(x,t)$ and $u_{i,k}(x,t)$ with $k = x, y, z$ to be equal. Dorozhkina and Semenov get an analytical solution of eqns. (5.4), (5.5) and (5.6) that depends on time-dependent scale lengths L_k, defined by the second moments of the electron distribution function:

$$L_k^2(t) = \frac{\int \mathrm{d}^3 x \int \mathrm{d}^3 v \, x_k^2 f_{e,i}}{\int \mathrm{d}^3 x \int \mathrm{d}^3 v \, f_{e,i}} = L_k^2 + c_{sk}^2 t^2, \qquad (5.9)$$

where c_{sk} with $k = x, y, z$ are the sound speeds. They are determined by initial values of electron and ion thermal velocities $v_{T(e,i),k} = \sqrt{k_B T_{(e,i),k}/m_{e,i}}$:

$$c_{sk}^2 = \frac{Z_i m_e v_{Te,k}(0)^2 + m_i v_{Ti,k}(0)^2}{Z_i m_e + m_i} = \frac{Z_i k_B T_e + k_B T_i}{Z_i m_e + m_i}. \qquad (5.10)$$

The following assumption is made for a self-similar solution for the electron and ion distribution function:

$$f_e = \frac{Z_i n_{i0}}{\sqrt{\pi}^3} \Pi_{k=1}^{3} \frac{1}{v_{Te,k}} \exp\left(-\frac{x_k^2}{L_k^2(t)}\right) \exp\left(-\frac{L_k^2(t)\,(v_k - u_{e,k}(x_k,t))^2}{L_k^2 \, v_{Te,k}^2}\right), \qquad (5.11)$$

$$f_i = \frac{n_{i0}}{\sqrt{\pi}^3} \Pi_{k=1}^{3} \frac{1}{v_{Ti,k}} \exp\left(-\frac{x_k^2}{L_k^2(t)}\right) \exp\left(-\frac{L_k^2(t)\,(v_k - u_{i,k}(x_k,t))^2}{L_k^2 \, v_{Ti,k}^2}\right). \qquad (5.12)$$

These equations can be used to calculate the electric field given by eq. (5.8). Inserting the distribution functions gives

$$eE_k = \frac{m_e m_i}{Z_i m_e + m_i} \frac{L_k^2 x_k}{(L_k^2 + c_{sk}^2 t^2)^2} \left(v_{Te,k}^2 - v_{Ti,k}^2\right) \quad (5.13)$$

$$\approx \frac{m_i}{Z_i} \frac{c_{sk}^2 L_k^2 x_k}{(L_k^2 + c_{sk}^2 t^2)^2}, \quad (5.14)$$

where $v_{Ti,k} \ll v_{Te,k} \approx c_{sk}$ has been assumed in the last step. Equation (5.14) shows that only the thermal pressures ($p \approx \nabla n_e \approx 1/L_k$) of all flow constituents add to the electric field while the mean velocities $\boldsymbol{u}_e$ and $\boldsymbol{u}_i$ cancel out. At the bunch periphery where the quasineutral approximation is not valid, the above equation is not valid either.

The electric field accelerates the flow velocity as

$$\frac{du_k}{dt} = Z_i e E_k / m_i, \quad (5.15)$$

and by remembering that $\frac{du_k}{dt} = \frac{\partial u_k}{\partial t} + u_k \frac{\partial u_k}{\partial x_k}$, the following differential equation is obtained:

$$\frac{\partial u_{i,k}}{\partial t} + u_{i,k} \frac{\partial u_{i,k}}{\partial x_k} = \frac{m_i}{Z_i} \frac{c_{sk}^2 L_k^2 x_k}{(L_k^2 + c_{sk}^2 t^2)^2}. \quad (5.16)$$

This equation can be solved by separation of variables $u_{i,k} = x_k g_k(t)$. Insertion results in

$$\frac{\partial g_k(t)}{\partial t} + g_k^2(t) = \frac{c_{sk}^2 L_k^2}{(L_k^2 + c_{sk}^2 t^2)^2}. \quad (5.17)$$

This Ricatti-ODE can be solved with $g(t) = c_{sk}^2 t/(L_k^2 + c_{sk}^2 t^2)$, yielding the following nonlinear differential equation for the ion mean velocity:

$$u_{i,k}(x_{i,k}, t) = \frac{c_{sk}^2 x_{i,k} t}{L_k^2 + c_{sk}^2 t^2}. \quad (5.18)$$

The expanding ion density can be calculated with $n_i = \int f_i \, d^3v$ and results in

$$\frac{n_i}{n_{i0}} = \Pi_{k=1}^3 \frac{L_k}{\sqrt{L_k^2 + c_{sk}^2 t^2}} \exp\left(-\frac{x_{i,k}^2}{L_k^2 + c_{sk}^2 t^2}\right). \quad (5.19)$$

This concludes the summary of the results for the expansion of a plasma plume obtained by Dorozhkina et al. [295, 296].

In the following the analytical solution by Dorozhkina et al. is compared to numerical solutions. In particular, the limiting velocity of the flow is determined in presence of a sharp cutoff in the density profile at low ion density. In order to obtain the velocity of an ion, the motion of a single ion within the self-consistent velocity field given by eqn. (5.18) is considered. The equations of motion are

$$\frac{dx_{ik}}{dt} = u_{ik} = x_{ik} \frac{B_k t}{1 + B_k t^2}, \qquad (5.20)$$

where $B_k = c_{sk}^2 / L_k^2$. Eqn. (5.20) can be integrated to obtain the ion trajectories as a function of their initial position x_{k0}

$$\int_{x_{ik0}}^{x_{ik}} \frac{dr_k}{r_k} = \int_0^t ds \, \frac{B_k s}{1 + B_k s^2}. \qquad (5.21)$$

The result for $x_{ik} \geq x_{ik0} > 0$, $x_{ik} \leq x_{ik0} < 0$, and $t \geq 0$ is

$$x_{ik} = x_{ik0} \sqrt{1 + B_k t^2}. \qquad (5.22)$$

Performing the coordinate transformation eqn. (5.22) in eqn. (5.18) the ion velocity as a function of the initial ion position x_{ik0} is obtained as

$$u_{ik}(x_{ik0}, t) = \frac{x_{ik0} B_k t}{\sqrt{1 + B_k t^2}} \rightarrow x_{ik0} \sqrt{B_k} \qquad (5.23)$$

for $t \rightarrow \infty$. The peak velocity of an ion depends on its initial position x_{ik0}, charge state Z_i, the masses m_i and m_e, and the thermal velocities v_{Tik} and v_{Tek}, respectively. For $x_{ik0} = 0$ an ion is in the center of the plasma plume with $u_{ik}(x_{ik0}, t) = 0$. The solution shows that there is a maximum velocity for each ion in the flow. The largest divergence angles θ_{ix} and θ_{iy} are

$$\theta_{ix} = \frac{x_{i0} \sqrt{B_x}}{z_{i0} \sqrt{B_z}}, \qquad \theta_{iy} = \frac{y_{i0} \sqrt{B_y}}{z_{i0} \sqrt{B_z}}. \qquad (5.24)$$

Now, the RCF signal for a spherical plasma expansion will be calculated. A schematic is shown in fig. 5.3. The plasma has initially the radius r_0. Later on, it is expanded to the radius r. At $z = z_{\text{RCF}}$ the detector is placed, sensitive to ions of the energy $E = E_{\text{RCF}}$. For a spherical plasma plume $B_k = B$ holds for $k = 1, 2, 3$. The fluid velocities are expressed as functions of position, dropping ion and electron labels,

$$u_k(x_k, x_{k0}) = x_{k0} \sqrt{B} \sqrt{1 - \frac{x_{k0}^2}{x_k^2}}. \qquad (5.25)$$

In what follows the ions are protons. It is assumed that protons start from initial positions

$$x_0 = r_0 \sin\theta \sin\phi, \qquad (5.26)$$
$$y_0 = r_0 \sin\theta \cos\phi, \qquad (5.27)$$
$$z_0 = r_0 \cos\theta. \qquad (5.28)$$

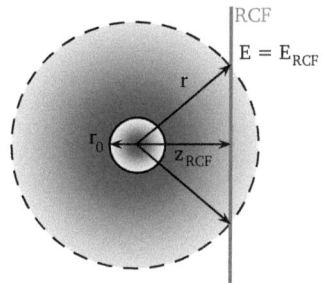

Fig. 5.3: Scheme of a spherical plasma expanding into an RCF detector.

Further it is assumed, that the target expansion is spherical meaning $x = r\sin\theta\sin\phi$, $y = r\sin\theta\cos\phi$, and $z = r\cos\theta$. Protons hitting the RCF detector have positions x, y, z_{RCF}. After some algebra the following relation is obtained for the energy of a proton hitting the RCF detector

$$E_{RCF} = \frac{1}{2} B m_p r_0^2 \left(1 - \frac{r_0^2}{r^2}\right), \qquad (5.29)$$

with $r^2 = x^2 + y^2 + z_{\text{RCF}}^2$. The initial density profile of the plasma plume is assumed to scale like

$$n_i = n_0 \exp\left(-\frac{r_0^2}{2L^2}\right), \qquad (5.30)$$

where $L_k = L$ for $k = 1, 2, 3$ denotes the density scale length. According to eq. (5.19) the density as a function of time, $n(r,t)$, scales as

$$n(r,t) = \frac{n_0}{(1+Bt^2)^{3/2}} \exp\left(-\frac{r^2}{L^2(1+Bt^2)}\right), \qquad (5.31)$$

where r is given by $r(t)^2 = r_0^2 (1 + B t^2)$ from eq. (5.22). The RCF measures particles with the energy E_{RCF}, hence the time t can be eliminated by the energy relation eq. (5.29) and $r(t)$. The solution is

$$t^2 = \frac{1}{B}\left(\frac{2r^2}{r_{min}^2}\left(1 - \sqrt{1 - \frac{r_{min}^2}{r^2}}\right) - 1\right), \qquad (5.32)$$

where $r_{min}^2 = 8 E_{\text{RCF}}/(Bm_p)$ and $r^2 = z_{\text{RCF}}^2 \left(\theta_x^2 + \theta_y^2 + 1\right)$.

Chapter 5. 3D proton expansion model

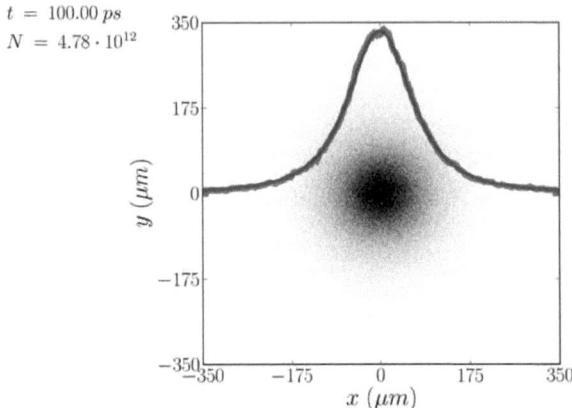

Fig. 5.4: Particle simulation of the expansion of a plasma plume in 3D as RCF would record it for protons. The parameters are: $n_0 = 6.65 \cdot 10^{28}\,\text{m}^{-3}$, $B = 10^{24}\,\text{s}^{-2}$ and $L = 10\,\mu\text{m}$. The plume contains $2.7 \cdot 10^{14}$ ions. The RCF is placed at $100\,\mu\text{m}$ distance from the center. Ions with $2\,\text{MeV}$ energy are recorded. The picture has been taken after $100\,\text{ps}$, at this point in time about $4.8 \cdot 10^{12}$ ions are recorded in the RCF. ( ) is a line-out at $y = 0$, (—) is the plot of eq. (5.31).

The energy E_{RCF} is arbitrary. For $r_{min} > r$, where r is the distance from the origin to a point in the detector, no protons are recorded. This implies that a ring like structure is generated in the detector. To give an example, eq. (5.31) is compared a numerical solution of the expansion. The density profile given in eq. (5.30) is represented by quasi-elements, which move according to the solutions given in eqs. (5.22) and (5.23). Quasi-particles with $2\,\text{MeV}$ are recorded in the RCF at $100\,\mu\text{m}$ distance from the center of the plasma plume. The initial electron temperature is $1\,\text{MeV}$, the scale length is $L = 10\,\mu\text{m}$ ($B = 10^{24}\,\text{s}^{-2}$). Figure 5.4 shows the result. 5×10^7 quasi-particles were sampled in 2000 time steps up to $100\,\text{ps}$. The background image shows the density distribution of the ions detected in the RCF. The red curve is a lineout at $y = 0$, smoothed by $\pm 7\,\mu\text{m}$ in y-direction and $\pm 3.5\,\mu\text{m}$ in x-direction. The blue curve is the plot of eq. (5.31). Both curves were normalized to 1. In a real experiment the scale length of the measured profile depends on the magnitude of B, hence it determines how the flow expands over time.

The important finding from this example is, that the imaging property of the expanding plasma is not an effect, that can be described by optical, geometrical methods (i.e., straight rays). Instead, the imaging depends on the velocity field of the protons. For this imaging no geometrical deformation of the target sheath is necessary. This new finding is in contrast to the model of the previous chapter, where a deformation is explicitly necessary. This is the major difference between the approach that takes into account the acceleration and an approach based on simple ray-tracing after the acceleration phase.

In order to determine how images of corrugations look, e.g. the grooves from a microstructured target rear side, an assumption of how perturbations propagate has to be made. They can propagate as plasma excitations. This would mean that they propagate with the local sound speed c_s on top of the flow. However, there is also the possibility that perturbations propagate kinetically within the expanding flow. In that case they do not trigger plasma excitations. It is assumed that the second is the case. Further, the perturbations shall be weak. Then the following equations of motion for those fluid elements that undergo perturbations have to be solved.

$$\frac{dx_k}{dt} = \frac{B_k\, x_{k0}\, t}{\sqrt{1+B_k t^2}} + \delta u_k \,, \tag{5.33}$$

where δu_k can be a function of the x_{k0}. The solutions of eq. (5.33) are

$$x_k = x_{k0}\sqrt{1+B_k t^2} + \delta u_k\, t \,. \tag{5.34}$$

Equation (5.34) can be solved for t. Since $t > 0$ has to hold it is

$$t = -\frac{\delta u_k\, x_k}{B_k x_{k0}^2 - \delta u_k^2} + \sqrt{\left(\frac{\delta u_k\, x_k}{B_k x_{k0}^2 - \delta u_k^2}\right)^2 + \frac{x_k^2 - x_{k0}^2}{B_k x_{k0}^2 - \delta u_k^2}} \,. \tag{5.35}$$

Equation (5.35) can be used to express the flow velocity as a function of the flow position as done in the previous example. It is, however, difficult to calculate the detector response analytically. The reason is that not all ions are effected by perturbations. Typically only those residing at the surface will be modestly perturbed due to the initial violation of quasi-neutrality at the surface.

5.2.2 The Charged Particle Transfer code - CPT

In the following a particle code is developed, that uses an approximation for the proton trajectories as a function of their initial positions and time. This code is called *Charged Particle Transfer - CPT*. With the code it is easy to propagate each fluid element separately and to accumulate its response in the detector. The analytical solutions for the expansion of a plasma plume are used as a guideline for the proton trajectories in the CPT. Experimental image data from chapter 4 serves to improve these approximations.

Prior to this it is demonstrated how CPT works. First, a two-dimensional (2D) PIC simulation with the PSC code is performed, showing the generation of the 2D equivalent of lines in RCF-stacks. The simulation has been carried out by Hartmut Ruhl. The simulation box $(x \times y \times z)$ for the PSC-simulation is $(1\,\mu\text{m} \times 40\,\mu\text{m} \times 15)\,\mu\text{m}$ large. The target in the simulation consists of a substrate of heavy material of thickness $d_\text{s} = 1\,\mu\text{m}$ with $m_\text{s} = 100\,m_\text{p}$, where m_p is the proton mass. The substrate is coated with a proton film of thickness $d_\text{f} = 0.1\,\mu\text{m}$. The whole target is singly ionized. The ion densities of the substrate and the film are both $n_\text{i} = 5 \cdot 10^{22}\,\text{cm}^{-3}$, about 1/10 solid density. A resolution of 30 cells per micron is used. The initial electron and ion temperatures are $T_\text{e} = 1\,\text{keV}$ and $T_\text{i} = 0\,\text{keV}$, respectively. The laser intensity is $I\lambda^2 = 5 \cdot 10^{18}\,\text{Wcm}^{-2}\mu\text{m}^2$. The laser beam has a line focus along x. The focal spot diameter along y at FWHM is $10\,\mu\text{m}$. The temporal envelope of the laser pulse is Gaussian with 200 fs at FWHM. The peak of the laser pulse is located $60\,\mu\text{m}$ in front of the foil at $t = 0$. The rear surface of the foil has sinusoidal micro-grooves with a wavelength of $2.5\,\mu\text{m}$ and an amplitude of $0.1\,\mu\text{m}$. The simulation runs for about 1 ps. The target in the simulation has been chosen such that in essential aspects it is as close to reality as possible but still feasible from the perspective of computational cost. To calculate the RCF detector response for the accelerated beam of protons obtained in the simulation a realistic detector model has been implemented in PSC that makes use of the stopping power characteristics of RCF for protons [254].

Figure 5.5 shows the proton phase space from the PSC simulation. Plot (a) gives the yz-plane. The plot shows the evolution of the proton flow with imprinted lines. The lines are flow perturbations generated by the electric field in the micro-grooves at the back surface of the target described in the simulation. An approximation for this electric field can be obtained from eq. (5.8). Neglecting the proton motion

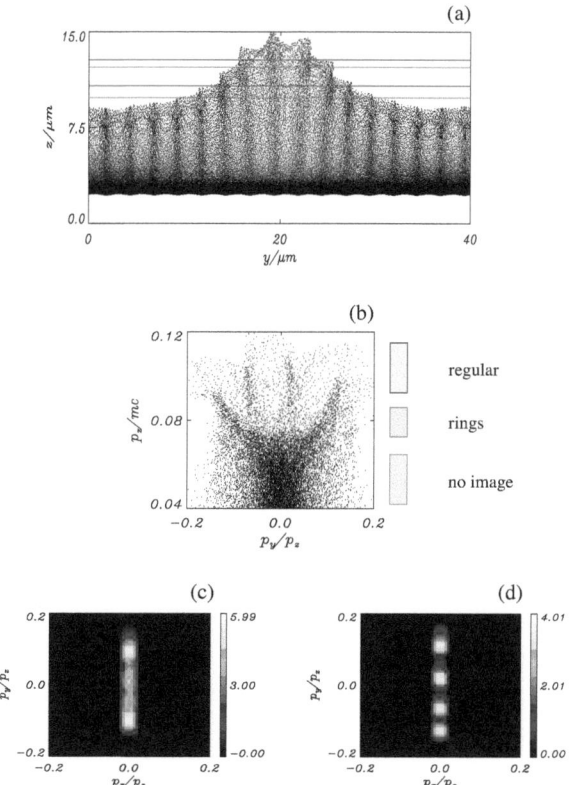

Fig. 5.5: Image classes obtained by a 2D PSC simulation. Plot (a) shows the yz-plane of the proton phase space at $t = 613\,\text{fs}$. Micro-corrugations are embedded into the flow. The red lines (ring region) define the momentum range $0.0775 \leq p_z/m_p c \leq 0.0825$ and the blue ones (no rings) $0.085 \leq p_z/m_p c \leq 0.09$. Both regions overlap in configuration space as is indicated by the blue and red lines in plot (a). Plot (b) shows the $\alpha_y p_z$-plane. The plot illustrates the different classes of images for various momenta p_z. The blue bar indicates the section of momentum space that generates regular groove patterns in the RCF-detector, the red bar shows the momentum range for which rings form, and the green bar represents the momentum range without images. Plot (c) shows the $\alpha_x \alpha_y$-plane for the momentum range bounded by the red bar in plot (b). Plot (d) shows the same for the momentum range bounded by the blue bar in (b). The asymmetry of the dots in plot (d) is due to the laser not hitting the symmetry axis of the target. This artifact shows how sensitive the image formation process is. Image courtesy of Hartmut Ruhl.

Chapter 5. 3D proton expansion model

and assuming a Gaussian electron distribution the field is $\mathbf{E} \approx k_\mathrm{B} T_\mathrm{e} \nabla n_\mathrm{e}/q\, n_\mathrm{e}$ (see eq. (2.49)). This field is normal to iso-density contour surfaces and micro-focuses the protons at early times. Plot (b) shows the $\alpha_\mathrm{y} p_z$-plane at 613 fs after the onset of the simulations. At this time the laser has already left the simulation box. The plot shows which structure information an RCF-detector will record at various p_z. The blue bar at the right side of the plot indicates the momentum range where regular groove patterns are to be expected in an RCF-stack detector. The red bar indicates the momentum range where ring structures with variable radii should appear. The green bar indicates the momentum range without any useful image information. Plots (c,d) show maps of the divergence angles θ_x and θ_y in color scales at 613 fs. While only two single bright dots are visible in plot (c) four individually resolved dots can be seen in plot (d). The two dots in plot (c) represent a ring structure in 2D. As can be seen rings occur at lower energies while they are absent at higher energies. Plot (a) of Fig. 5.5 shows that the divergence angles of the flow are largest in the center, where laser-heating of electrons is strongest, and that they are smallest at peripheral target locations. The rings are composed of protons originating from target locations between large and small divergence angles. In case the detector has a broad momentum acceptance of $0.06 \leq p_z \leq 0.08$ ($1.67\,\mathrm{MeV} \leq E_z \leq 3\,\mathrm{MeV}$) centered around $p_z = 0.07$ ($E_z = 2.3\,\mathrm{MeV}$) it will record a ring with a dark dot in the center from protons with small momenta p_z. From plot (a) is is determined that those come from the peripheral target locations. The flow perturbations widen as the flow expands. However, they cannot overcome the overall expansion of the flow.

In what follows it is tried to construct an analytic approximation to the experimentally measured proton flow. The RCF-images of micro-perturbations embedded in the proton flow will be used as a diagnostic. At an early stage of the proton acceleration a strong, tightly bounded sheath field exists at the back surface of the irradiated target. Times are always measured from the time when the first fast electron hits the back surface of the target. At this time proton acceleration is very rapid but yet there has been not enough time for them to move significantly. During this phase quasi-neutrality in the flow is strongly violated. However, since the protons had no chance to move the proton distribution in configuration space can be easily predicted, since they have not moved in momentum space as well.
At a later stage hydrodynamic expansion of the flow sets in and the plasma becomes quasi-neutral. This process is very rapid. Hydrodynamic expansion destroys the sharp density gradients initially present at the rear surface and hence tends to extend and weaken the sheath electric field. During this phase eq. (2.49) applies.

The hydrodynamic expansion of the flow, however, is essential for the generation of images of rear surface structures in the RCF. Hence, an analytic approximation to the experimental proton flow can be constructed by two essential assumptions. The first assumption is the shape of the proton distribution function in the sheath field close to the rear target surface at very early times. The second one is a plausible approximation to the transfer function. The transfer function expands the initial proton distribution at the rear surface. The analytic model shows that for the hydrodynamic expansion phase where the assumption of quasi-neutrality is good such functions exists, in the form of eq. (5.25).

5.2.3 Transfer function derived from a fluid approach

The simplest approach to derive a transfer function for the flow expansion is a quasi-neutral, isothermal expansion, described in sec. 2.4.1. The electric field (eq. (2.49)) is $\mathbf{E} = -k_B T_e/e \, \nabla n_e/n_e$. By inserting this field in the ion component of eq. (2.46) and by observing that $T_i \ll T_e$, the force acting on the ions is $\dot{\mathbf{p}}_i = -k_B T_e \nabla n_e/n_e$. A reasonable assumption for the electron density profile is a Gaussian in transverse directions x, y and an exponentially decaying profile in longitudinal direction z:

$$n_e = n_0 \exp\left(-\frac{(x-x_0)^2 + (y-y_0)^2}{2L^2} - \frac{z-z_0}{L_z}\right). \tag{5.36}$$

Now the envelope functions $p_{x,y,z}$, that describe the expanding proton flow are derived. On top of the expanding flow momentum perturbations propagate, induced by the sinusoidally shaped surface. The following equations give details of this model

$$\frac{dp_{yi}}{dt} = C_y(y_i - y_0), \quad \frac{dy_i}{dt} = \frac{p_{yi}}{m_p}, \quad p_{yi}(0) = S_y(y_i(0) - y_0) + \delta p_{yi}, \tag{5.37}$$

where $C_y = k_B T_e/L^2$ and $\delta p_{yi} = A_y \sin(k_y y_i(0))$. The longitudinal flow expansion is obtained as

$$\frac{dp_{zi}}{dt} = C_z, \quad \frac{dz_i}{dt} = \frac{p_{zi}}{m_p}, \quad p_{yi}(0) = S_z(z_i(0) - z_0), \tag{5.38}$$

with $C_z = k_B T_e/L_z$. The quantities $p_{yi}(0)$, $p_{zi}(0)$, $y_i(0)$, and $z_i(0)$ describe the proton distribution at $t = 0$. Equation (5.37)(left) means that the lateral momentum spread of the protons increases faster the further away a proton is from the center of the target as is observed in the simulation. The quantities y_i and y_0 represent the lateral position of a proton and the target center, respectively. The quantity C_y determines how fast the lateral flow expands and is proportional to the electron temperature and the inverse scale length. S_y represents the initial inclination of the momentum envelope, that is determined by the early rapid expansion of the flow in the sheath. The quantities C_z and S_z mean the same for the longitudinal expansion of the flow. The parameters z_i and z_0 represent the longitudinal position of a proton in the flow and the position of the rear target surface, respectively. Equations (5.37) and (5.38) can be solved analytically as shown in ref. [88].

5.2.4 Transfer functions for TNSA-protons

The combination of an initial spatially bounded proton distribution function with a transfer function for the expansion phase naturally recovers energy cutoffs in the spectrum of accelerated protons at low and high energies. However, the analytical solution given by the transfer functions from above cannot describe the situation of a proton film attached to a heavy substrate in 3D. Hence, a different approach is adopted. The transfer function for the flow is constructed with the help of experimental data. The early proton distribution function in the thin proton film is set up as a Gaussian, motivated by the energy-resolved source size measurements in sec. 4.1.4:

$$f_p(\mathbf{x}, \mathbf{p}, 0) = N_p\, n_p(\mathbf{x}) \exp\left(-\frac{[\mathbf{p} - \mathbf{P}(\mathbf{x})]^2}{2 m_p k_B T_p}\right) , \quad (5.39)$$

$$n_p(\mathbf{x}) = n_0 \exp\left\{-\frac{z - z_0}{L_0 + L\, h(x, y)}\right\} , \quad (5.40)$$

$$h(x, y) = \exp\left\{-\frac{x^2 + y^2}{r_0^2}\right\} . \quad (5.41)$$

The normalization factor N_p is obtained by integrating over the total volume of the film attached to the substrate and equating the result to the total number of protons found there. The parameter L represents the scale length of rapidly accelerated protons opposite to the location where the peak laser intensity is found on the front side of the target. The parameter r_0 is the effective radius of this super-heated area. In addition to super-heating there is fast lateral electron transport and electron recirculation. To take account of this effect the effective parameter L_0 has been introduced. To model the boundedness of the early proton film

$$z \leq z_0 + (L_0 + L\, h(x, y)) \ln \frac{n_0}{n_p} , \quad (5.42)$$

is required, where n_p is the cutoff density which is adapted to best match the experiment. The value n_0 represents the proton background density. The parameter z_0 represents the location of the plane of the proton sheath. For a rear surface with sinusoidal grooves an adequate choice for the early beam perturbations is

$$P_x = A_x\, m_p c\, \sin(k_x x)\,, \qquad (5.43)$$

$$P_y = A_y\, m_p c\, \sin(k_y y)\,, \qquad (5.44)$$

$$P_z = A_z\, m_p c\, (z - z_0) \qquad (5.45)$$

since sheath acceleration is essentially normal to the density iso-contour surfaces. There are ambiguities about the early form of the proton distribution function and the rate at which the beam expands. It is not possible at present to measure the momentum of protons as a function of their position in the flow. Such a measurement would directly determine the transfer function. Hence, it is tried to infer this information indirectly from the RIS data. The more experimental data are available, the better the approximation can become.

The transfer function maps the positions of the protons to their respective momenta. This is denoted by $\mathbf{g}$. Then the following equations of motion for the proton flow, represented by the phase space coordinates $\mathbf{x}_i$ and $\mathbf{p}_i$, have to be solved

$$\mathbf{p}_i(t) = \mathbf{g}\,[\mathbf{x}_i(t)] + \delta\mathbf{p}\,[\mathbf{x}_i(0)] \qquad \mathbf{x}_i(t) = \mathbf{x}_i(0) + \frac{1}{m_p}\int_0^t d\tau\, \mathbf{p}_i(\tau)\,, \qquad (5.46)$$

where $\delta\mathbf{p}\,[\mathbf{x}_i(0)] = \mathbf{p}_i(0) - \mathbf{g}\,[\mathbf{x}_i(0)]$. The function $\delta\mathbf{p}$ defines the initial beam perturbations embedded into the flow.

A sufficiently accurate approximation for $g_{x,y}$ is motivated by eqs. (5.37) and the assumption of a transverse Bell-shaped hot electron sheath that results in a linearly increasing force in z-direction.

The particular choice of g_z has been motivated by the observation of the momentum distribution p_z versus z in the PSC simulation. Figure 5.6 (left) shows this distribution from such a simulation. The color-coding represents the number of the protons, with red color being the maximum number and darker colors representing lower numbers. The right side of the figure shows a logarithmic fit to the data, that well represents the bulk of the proton distribution. At the tip, encircled by the green ellipse, the non-neutrality results in an enhanced acceleration compared to the bulk. Hence the p_z deviates from the logarithmic curve. The proton number is quite low and they are neglected in the CPT. These observations finally lead to the following transfer functions:

$$g_x = C_r \, m_p c \, \frac{(x - x_0)(z - z_0)}{r_0^2} \,, \tag{5.47}$$

$$g_y = C_r \, m_p c \, \frac{(y - y_0)(z - z_0)}{r_0^2} \,, \tag{5.48}$$

$$g_z = C_z \, m_p c \, \ln\left(\frac{z}{z_0}\right) \,, \qquad z_0 \leq z \leq z_1 \,. \tag{5.49}$$

The parameters C_r, C_z, $\mathbf{A}$, L_0, L, n_p, T_p, r_0, and z_0 represent free parameters chosen such that RCF-data such as spectrum, energy-resolved opening angle and source sizes, emittance and transverse beam profile can be reproduced. The particular choice of the envelope functions is obtained from measurements, simulation, and the requirement that the flow expands mostly normal to iso-contour surfaces of the density.

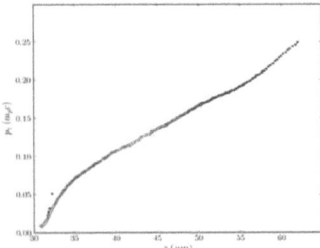 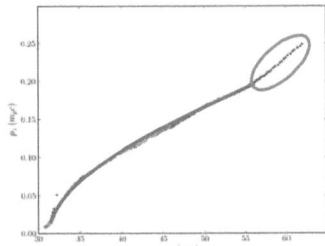

Fig. 5.6: Momentum distribution p_z versus position z of the protons from a PSC simulation. The color-coding represents the number of the protons, with red color being the maximum number and darker colors representing lower numbers. The right side of the figure shows a logarithmic fit to the data, the green ellipse encircles super-accelerated protons in the field region at the tip.

5.2.5 Numerical implementation

The CPT code solves eqs. (5.46) by the midpoint method [297]. Quasi-particles representing the real protons are used in the calculation. Initially, the particles are distributed with random distances around the shape given by eqs. (5.39). The code is divided in two parts: the first part does the calculation and data pre-selection. It has been written in C for fast execution speed. The second part then plots the data and is written in Python with the `matplotlib`-extension.

5.3 Reconstruction of experimental data

The CPT will now be used to fully reconstruct a proton beam accelerated at the Z-Petawatt. The laser was focused to $11\,\mu$m FWHM with an energy of 35 J onto a $25\,\mu$m thick, micro-structured gold foil. The rear side had sinusoidal grooves with a $10\,\mu$m line distance. Assuming 45 % of the energy in the focal spot, the laser intensity was $I > 1.6 \times 10^{19}$ W/cm^2. The pulse length is not exactly known, therefore it is assumed to be 1 ps. Hence the acceleration time, according to eq. (2.71), is $t_{acc} = 2.2$ ps. This time span is used in the CPT simulation. The energy-resolved opening angle, source size, transverse phase space and spectrum were given by the RIS data. The experimental spectrum could be well fit by eq. (3.14) with $N_0 = 5.07 \times 10^{11}$ and $k_B T = 8.9$ MeV. The other parameters were already given in sections 4.1.3 - 4.1.5.

This set of parameters will now be reconstructed in the CPT. The initial distribution has been set-up according to eqs. (5.39). The calculation has been set up to sample 5×10^7 quasi-particles. The initial FWHM radius is taken from the source size measurement (figure 4.4) to $r_0 = 90\,\mu$m. The position $z_0 = 25\,\mu$m is the rear side of the target, measured from the target front side. The scale length L is the Debye length $\lambda_D = 0.73\,\mu$m of the plasma. The initial ion temperature $k_B T_p = 20$ eV is taken from PIC simulations and emittance measurements [86]. This value determines the contrast of the imprinted surface grooves in the RCF. The surface grooves are modeled as perturbations of the initial momenta, set-up according to eqs. (5.43) and (5.44) with $A_x = 0$, $A_y = 3 \times 10^{-4}$ and $k_y = 2\pi/10\,\mu$m^{-1}.

The rest of the parameters are determined by fitting the experimental data. The most sensitive parameters are the particle spectrum, determined by N_p, n_p, C_z and the energy-dependent opening angle, determined by A_z, C_r, C_z. The total density $N_p = 4 \times 10^{22}$ cm^{-3} determines the total number of the protons, i.e., the height of the spectrum. The slope of the spectrum is very sensitive to the cut-off density n_p in combination with the acceleration strength C_z. The best fit could be obtained by

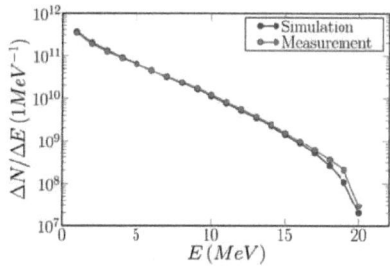

Fig. 5.7: Spectrum from CPT simulation compared to the measured spectrum.

$n_p = 10^{-3} N_p$ and $C_z = 0.139$. The resulting particle spectrum (fig. 5.7) from the CPT simulation ($-\circ-$) agrees very well with the measured spectrum ($-\circ-$).

The energy-resolved opening angle can be fit by adjusting C_z, C_r, A_z. The acceleration strength C_z determines the maximum energy of the beam, the opening angle is determined by C_r. Below some certain energy, the opening angle do not increase anymore, but stays constant or decreases. The position of this inflection point is controlled by the proper choice of A_z.

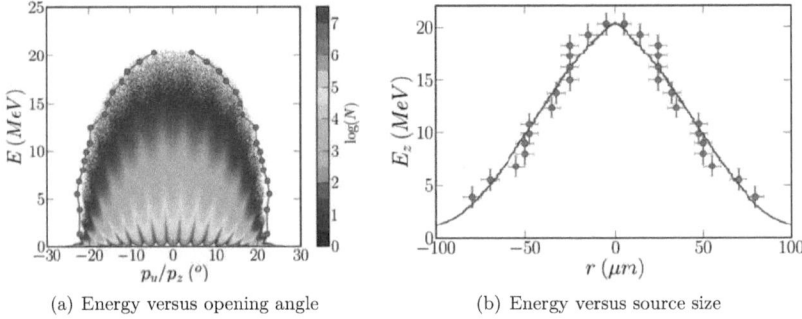

(a) Energy versus opening angle   (b) Energy versus source size

Fig. 5.8: Comparison of CPT-simulation versus measured data. a) shows the good agreement of the energy-resolved opening angles from the CPT (color coding represents number of quasi-particles) and the measured data ($-\circ-$). b) shows the energy-resolved source size. The CPT simulation ($---$) fits the measured data ($\circ$) very well.

The best fit was obtained for $A_z = 2 \times 10^3$, $C_z = 0.139$ and $C_r = 0.13$, respectively. The comparison between simulated values (the color-coding represents the number of quasi-particles, with maximum $N_0 = 5 \times 10^7$) and the measured data ($-\circ-$) are shown in figure 5.8a. The opening angle α is determined in the CPT simulation by the ratio of the transverse momentum p_y by the longitudinal momentum p_z. The kinetic energy is given by $E_{\text{kin}} = \boldsymbol{p}^2/2m_p$. An excellent agreement is found for the simulated and measured beam envelopes. In addition to that, the CPT simulation shows the appearance of lines, originating from the initial momentum perturbation. The energy-resolved source size is shown in figure 5.8b. The CPT simulation ($—$) fits the measured data ($\circ$) very well. This plot shows, that the width and shape of the initial distribution $h(x, y)$ in eq. (5.39) is directly mapped to the shape of the transverse particle energy distribution.

The last part of the reconstruction is the comparison between the simulated and measured RCF images. Figure 5.9(right) shows the simulated RCF image of 12 MeV protons in comparison to the measured beam profile on the left side. There is a good agreement between the two images. The lines in the measured data are not as perfect as those ones from the simulation, most likely due to a not-so-perfectly Gaussian shaped initial sheath in reality.

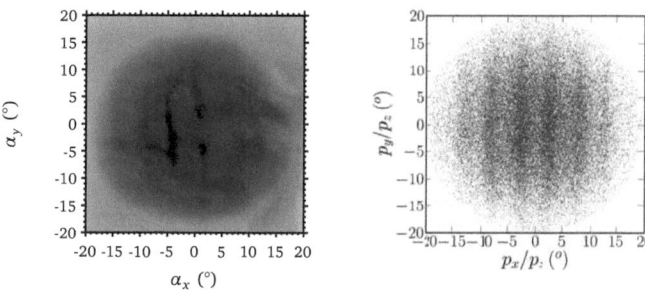

Fig. 5.9: Comparison of RCF images from measured data (left) versus CPT-simulation (right).

In conclusion, the CPT model can be used to reconstruct the acceleration phase of TNSA-protons. The flow expansion is determined by the initial shape of the particle distribution and a proper choice of the transfer functions. By fitting the energy spectrum to the data, an exponentially increasing hot electron density of $4 \times 10^{19}\,\text{cm}^{-3}$ was determined. The total density of $N_p = 4 \times 10^{22}\,\text{cm}^{-3}$ from the CPT-fit is close to the density of all protons in an area with radius of 0.5 FWHM and 12 Å thickness (see end of section 4.1.1), that is $n_0 = 6.5 \times 10^{22}\,\text{cm}^{-3}$. It is worth noting, that the insertion of the cut-off density $n_p = 4 \times 10^{19}\,\text{cm}^{-3}$ and the hot electron temperature from the laser's ponderomotive potential (eq. (2.28)) of $k_B T_e = 1.3\,\text{MeV}$ in the scaling given by eq. (2.70) results in a maximum energy of 20.5 MeV. This value agrees very well with the maximum energy determined in the experiment. Since these two independent models (CPT and the plasma expansion model) lead to the same density, it is very likely that the sheath density at the rear side was $n_{\text{hot}} = 4 \times 10^{19}\,\text{cm}^{-3}$ and the initial hot electron temperature was $k_B T_e = 1.3\,\text{MeV}$. A very recent publication by Antici et al. has determined the same density with a different technique for similar laser parameters [298].

The temperature of the expanding plasma is not explicitly included in the transfer functions used in the CPT model, in contradiction to the simplistic model in sec. 5.2.3. The relation to a one-dimensional fluid expansion model is given by the

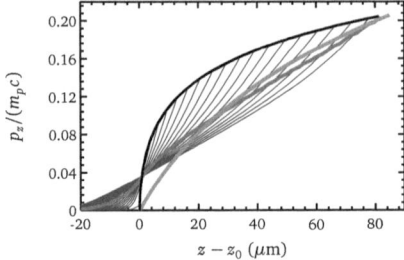

Fig. 5.10: Simulation with the fluid code from section 2.4.1 with the plasma parameters as obtained from the CPT fit. The proton momentum p_z versus z (—) changes strongly as the plasma expands. The front velocity versus space (eqs. (2.68) and (2.68)) is given by (—). The time-averaged momentum versus position (—) is very close to the transfer function from the CPT as given by eq. (5.49) (—).

observation of the momentum distribution versus space, for different times during the expansion. Figure 5.10 shows the result of a simulation with the fluid code from section 2.4.1 with the plasma parameters as obtained from the CPT fit. The simulation time was 2.2 ps. The proton momentum p_z versus z (—) changes strongly as the plasma expands. The front velocity versus space (eqs. (2.68) and (2.68)) is given by (—). The time-averaged momentum versus position (—) is very close to the transfer function from the CPT as given by eq. (5.49) (—). Hence, the transfer function used in the effective code CPT represents the average momentum increase of a proton as its changes its position during the acceleration.

5.4 Expansion dynamics

The analytical investigation of the flow expansion in section 5.2.1 and the PIC simulation in fig. 5.5 have shown, that the imaging properties of the proton beam are dictated by the expansion in momentum space. The grooves at the rear side lead to a perturbation in momentum space, that expands with the flow. The divergent proton acceleration corresponds to a shear and collimation of the ellipse in transverse phase space. Fig. 5.11(upper row) shows the history of the divergence angle p_y/p_z versus initial position r for protons with four different final energies. The black points show protons having a final energy of 2 MeV. Initially (left image) the divergence angles are sinusoidally distributed in space with equal amplitudes. An RCF records the divergence angles, as given by eqs. (3.11). Therefore the RCF measurement represents the projection of the phase space on the p_y/p_z-axis. For the left image, the RCF would record a homogeneous distribution of ions. As the flow

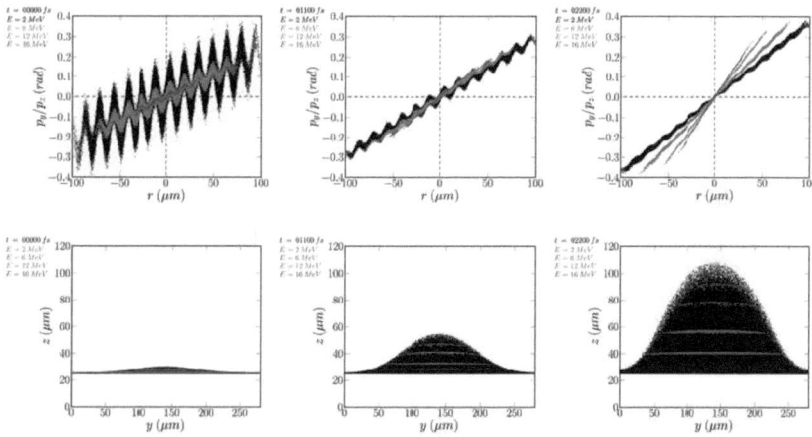

Fig. 5.11: Phase space (upper row) and spatial proton distribution in the CPT model. The three columns correspond to three different points in time at $t = 0, 1.1, 2.2\,\text{ps}$ during the acceleration, respectively.

expands and accelerates, the phase space distribution shears and the sine-pattern becomes smaller and more localized. The projection of the phase space to the p_y/p_z-axis results in a distribution that has regions with more particles, hence the RCF image becomes darker and lines becomes visible.

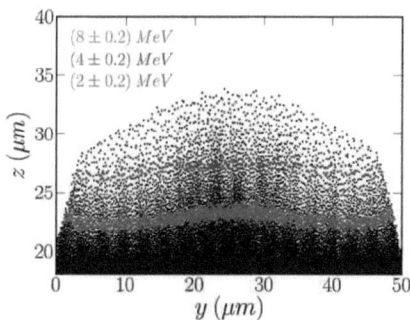

Fig. 5.12: Proton distribution from the PIC simulation.

At the same time, the protons expand in space. The lower row of fig. 5.11 shows the (y, z)-coordinates of the particles. Some particles are color-coded according to their final energy at $t = 2.2\,\text{ps}$. The beam is highly laminar, i.e., particles do not overtake each other. Even more important, the particle distribution stays very flat during the acceleration stage. This is in contradiction to expansion models that need a curved shape in space to explain the divergence of the beam, as e.g. the SABRINA model or the model from ref. [243, 290]. The reason is, that both the SABRINA model and the model in ref. [290] were used to explain the shape of the ions *after* the acceleration stage and not during the acceleration. As shown with the CPT, the distribution of protons with some certain energy is still very flat at the end of the acceleration. However, the large transverse momentum p_y for protons far off the center leads to a lower longitudinal momentum p_z, compared to ions of the same total kinetic energy $E = (p_x^2 + p_y^2 + p_z^2)/2m_p$ at the center where $p_x = p_y = 0$. This naturally leads to a curved ion front during the ballistic expansion. The p_y-momentum increase is linear with y. Since the velocity vector is pointing in the direction of the gradient of the ion front, the resulting ion front will be parabolically shaped for late times, just as found by ref. [243, 290].

These arguments are strongly supported by the observation of the ion distribution in space in the 2D-PIC simulation from section 2.4.2. Fig. 5.12 shows such a distribution at the end of the simulation run at about 1 ps. Particle positions according to three energies are marked. The particle distribution for some certain energy is again flat. For later times, which corresponds to larger values of z, the flat distribution starts to bend and becomes parabolic (e.g. the red particles in the figure) as expected from the CPT analysis. The sharp kink in the distribution at the left and right edges ($y \leq 5\,\mu$m and $y \geq 45\,\mu$m, for $z \leq 28\,\mu$m) is due to the finite bounding box, that could not be made larger because of the limited computer memory and computation time. The real envelope of the particles would decay as a Gaussian, in analogy to the CPT model (cf. fig. 5.11, lower row, center).

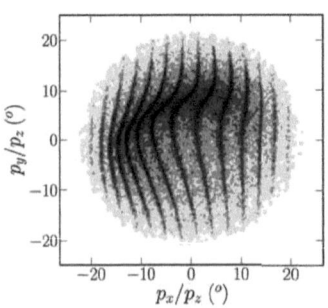

Fig. 5.13: RCF image of 8 MeV protons from CPT, with an additional elliptical perturbation.

The SABRINA model presented above was used to reproduce the shape of the sheath after the acceleration, i.e., when the ion fronts are already curved. A close correlation to the laser focal spot shape could be shown. The analysis in sec. 5.1.1 has shown, that the imprint of the laser focal spot leads to a perturbation of the initial conditions only. These perturbations are in momentum space, due to a non-Gaussian hot electron sheath prior to the ion acceleration, similar to the grooves at the target surface. Unfortunately, this case cannot be modeled with the current version of the CPT since the numerical implementation needs an analytical form of the perturbation like the sinusoidally shaped initial perturbation and not an arbitrary one as used in SABRINA. For an estimate, an elliptically shaped, rotated off-axis perturbation has been added in the CPT in order to mimic the effect of a laser line-focus. All other parameters have been kept unchanged compared to the previous section. Figure 5.13 shows the resulting RCF image for 8 MeV protons. The elliptical perturbation leads to a bending of the lines in the beam profile, while the background of the image is still round. The image qualitatively agrees to the observations from section 4.2. The additional elliptical perturbation reflects irregularities in the initial hot electron sheath. The later expansion, however, is still governed by the transfer functions for a smooth, regular Gaussian beam. Hence the other beam parameters as e.g. the particle spectrum (fig. 5.7) are unchanged, as found by the analysis of the experimental results with SABRINA.

5.5 Summary

In summary, two different models have been presented in order to understand the expansion properties and imaging characteristics of TNSA protons. The first model, SABRINA, can be applied to reproduce the distribution of protons with a discrete energy at the end of the acceleration process. The model was used to explain the laser beam profile impression and target thickness impact on laser-accelerated protons. Further, a discussion of the electron transport prior to the acceleration lead to the conclusion, that for thick targets ($d \approx 50\,\mu$m) the electron transport is dominated by small-angle collisions. For thin targets ($d < 15\,\mu$m) small-angle collisions play a minor role. The observed large source sizes of the protons are most likely the result of re-circulating electrons.

The physics of the expanding plasma has been investigated by considering the expansion of a Gaussian plasma plume. The important finding from this investigation is, that the imaging property of the expanding plasma is not an effect that can be described by optical, geometrical methods (i.e., straight rays). Instead, the imaging depends on the velocity field of the protons during the acceleration. For this imaging no geometrical deformation of the target sheath is necessary, which is the major difference to the SABRINA model. The flow expansion can be characterized by a transfer function that maps the particle momenta to their respective positions. Based on these observations, the CPT code has been developed and successfully applied to a full reconstruction of the proton beam accelerated at the Z-PW laser at SNL. Based on the CPT reconstruction, the initial hot electron sheath could be reconstructed to be of a Gaussian shape with $90\,\mu$m FWHM in transverse direction and exponentially decaying longitudinally. The hot electron density could be determined to be $4 \times 10^{19}\,\mathrm{cm}^{-3}$. The hot electron temperature was deduced by the plasma expansion model to be of 1.3 MeV, equal to the laser ponderomotive potential.

2D-PIC simulations further support the expansion dynamics from the CPT model. In particular, the particle distribution stays flat in space during the acceleration, i.e. no strong geometrical deformation is developed. The transverse momentum increases the further off the center a particle is. This leads to an divergent flow and a curved ion front for very late times after the acceleration phase. It was argued, that irregularities of the hot electron sheath are reflected in perturbations of the initial momenta, while the spatial distribution of the particles stays unchanged. The expansion is governed by the transfer functions for a smooth, regular Gaussian beam, demonstrating the similarity of TNSA for all laser systems.

Chapter 6
Outlook

Laser-accelerated ion beams have the potential to be used in a large variety of applications. The beams can be used as a diagnostic tool (e.g. proton radiography of transient processes [25–27]) and they could have applications as compact particle accelerators [31–34], for the creation of high-energy density (HED) matter [29, 30], as a driver for neutron production [35, 36], for radioisotope generation [37–40], for table-top nuclear physics [41], for the generation of intense K_α x-rays [42], for Inertial Fusion Energy as in the case of Proton Fast Ignition [43, 44] or even for medical applications as a compact radiotherapy system for tumor treatment [45–48].

Out of this huge field, one application of great relevance in basic plasma research will be discussed in more detail, namely the generation of high-energy density matter by laser-accelerated protons and its diagnostics by spectrally resolved x-ray scattering. Before that, the following section presents some ideas for a further optimization and control of laser-accelerated protons.

6.1 Further optimization and control

The experimental and theoretical investigations in the previous chapters have shown, that laser-accelerated ions basically resemble a quasi-neutral plasma expansion with a charge-separation at the ion front. Therefore, in order to increase the maximum energy, the hot electron temperature could be increased by increasing the laser power. Published experimental data suggest that with current laser technology a high-energy PW laser with a peak intensity above 10^{21} W/cm^2 is necessary for the acceleration of protons with high energies. Figure 6.1 shows the scaling of maximum energy with laser intensity for a 25 μm thick metal foil. Shown are data from LULI-100 TW (•) [15], VULCAN-PW (•) [16], the NOVA Petawatt (•) [21] and

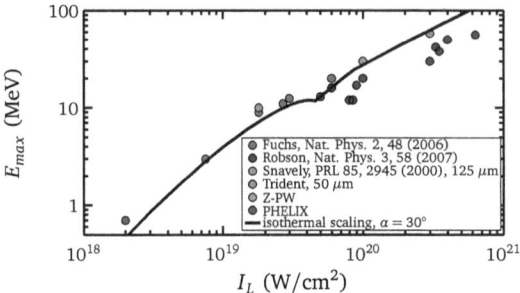

Fig. 6.1: Scaling of maximum energy with laser intensity, for a 25 μm thick metal foil. Shown are data from LULI-100 TW (•) [15], VULCAN-PW (•) [16], the NOVA Petawatt (•) [21] and data obtained in this thesis, compared to the scaling explained in section 4.1.2 assuming an electron divergence angle of $\alpha = 30°$.

data obtained in this thesis, respectively. The data reasonably well scale with the isothermal expansion model explained in section 4.1.2 assuming an electron divergence angle of $\alpha = 30°$. It should be noted however, that the plotted data are for lasers with pulse duration about 500 ps and targets with a 25 μm thickness or more. Data obtained at laser systems with much less pulse duration and/or much thinner targets deviate from the scaling, but no scaling relation for ultrashort laser pulses has been published so far.

The requirement of a high-energy laser is partially due to the limited pulse contrast on the order of 10^{-6} in present-days laser systems. The pre-pulse level (temporal contrast) can be significantly reduced with double plasma mirrors [299, 300]. A plasma mirror becomes reflective only when a certain intensity threshold to spark a plasma is reached. This allows the use of ultra-thin target foils and therefore very efficient proton acceleration [116, 117]. However, the necessary double plasma mirror configuration greatly reduces the energy by about a factor of five or more, hence the requirement of a high laser energy still remains.

The option of using circularly polarized laser beams instead of linearly polarized ones is currently investigated theoretically [122–125]. A circularly polarized laser pulse might generate high-energy, low-divergence ion beams with a non-exponential spectrum. Due to the circular polarization, the electrons do not gain a very high energy in longitudinal direction. The electrons are pushed forward by the laser's ponderomotive force, dragging the ions behind them. This allows for the acceleration of the foil as a whole. However, this scheme requires high-intensity pulses with very high pulse contrast as well, making it very difficult to realize experimentally.

Furthermore, a circularly polarized, ultra-high intensity ($I > 6.8 \times 10^{22}\,\text{W/cm}^2$) beam could be used in combination with a high-density ($n_e = 1.5 \times 10^{21}\,\text{cm}^{-3}$) gas jet, in order to accelerate protons in the electric field created by an electron bubble in the laser wake field [301]. For even higher intensities ($I_L > 1.37 \times 10^{23}\,\text{W/cm}^2$) and ultra-high contrast, the radiation pressure can directly accelerate ions to GeV energies from a solid target [194]. Nonetheless it is not clear whether or not the efficiency and beam quality of beams generated by these mechanisms is equal or better than TNSA-beams. Additionally, the experimental realization might only become realizable with the next generation of high-energy, high-intensity laser systems.

There are some other options to increase the efficiency of TNSA with the existing generation of laser systems. In order to increase the hot electron temperature, a confinement of the pre-plasma at the front side and the region of ion-acceleration at the rear side is proposed. The cone-shaped targets have been proven to enhance the conversion efficiency from laser energy to hot electron temperature. Further investigations for more efficient TNSA should be done with these cones, in order to get a better scaling with the cone dimensions and to find the optimum geometrical shape. Additionally, at the tip of the cone, a Gaussian shaped foil could be placed. The Gaussian shaped foil could lead to a reduction of the proton opening angle and, due to the confinement of the expanding plasma for early times, the adiabatic cooling due to the expansion would be slower, increasing the maximum energy as well. An energy-enhancement of the protons has been already observed in 2D-PIC simulations in refs. [302,303]. Up to 30 % higher maximum energy could be obtained by a strongly curved foil [302].

Furthermore, a curved rear surface could be used for a better collimation of the beam, that is necessary for applications. Therefore only moderately curved foils

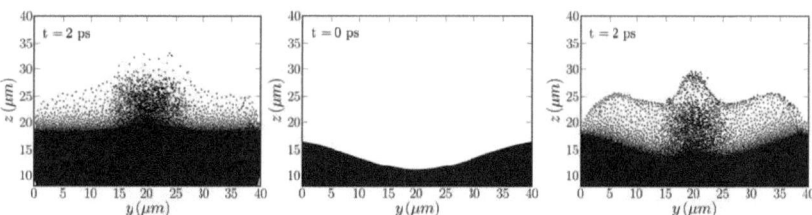

Fig. 6.2: Comparison of the proton distribution for a flat foil (left image) at $t = 2\,\text{ps}$ and a Gaussian shaped foil (center and right images). The curved target surface leads to a collimation of the beam.

should be used. While the experimental proof has still to be done, the effect of a Gaussian shaped foil on the proton beam expansion has been studied with the PSC. Figure 6.2 shows the (y,z)-distribution of protons in a PIC model-simulation with a Gaussian shaped foil, compared to a simulation with a flat foil and otherwise identical parameters. The target consisted of protons and electrons only. The laser intensity was very moderate, with $I = 10^{18}\,\text{W/cm}^2$. Due to the limited computation box size, the Gaussian shape has been set up with a FWHM of $15\,\mu\text{m}$ and a height of $6.5\,\mu\text{m}$. In contrast to the diverging beam observed in the simulation with a flat foil (left image), the protons are clearly converging at first (right image). Hence the intrinsic divergence due to the initial hot electron sheath could be compensated. In contrast to the studies in refs. [302, 303] the curvature is still weak, hence the maximum energy increase is only 3 % compared to the flat foil simulation. Due to the extraordinary long computation time systematic studies with different Gaussian shapes could not be performed. However, the simulation has shown that Gaussian shaped foils are a possible way to decrease the large opening angle, while at the same time the maximum energy can be slightly enhanced.

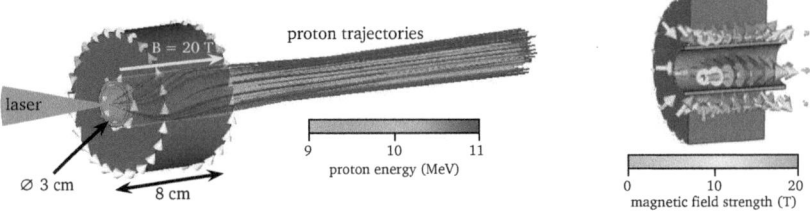

Fig. 6.3: Current-driven solenoid for the collimation of 10 MeV protons. The coil produces a homogeneous magnetic field of $B = 20\,\text{T}$, collimating the protons to a parallel beam of 30 mm diameter.

Another method for the reduction of the intrinsic divergence is the use of an externally applied field. This could be either a laser-driven microlens [81, 82] with its drawbacks that it needs two synchronized laser beams and that is has to be replaced after each shot; or just a magnetic field as demonstrated in section 4.4.2. However, the strong divergence of the beam either requires very large aperture ion optics (several ten cm) or very strong magnetic fields ($B \gg 1\,\text{T}$) close to the source. Neglecting the laser hardware, one big advantage of TNSA-protons is the compactness of the acceleration device. Hence placing a large-diameter ion optics device behind the target is not very attractive. Strong magnetic fields in turn can be generated by a compact, pulsed power device, using similar or even the same capacitors like those used to trigger the flash lamps of the laser system. The magnetic field should act like an ideal lens in optics, collimating the beam originating from a point-like source. An adequate magnetic field that offers a good imaging quality even for large angles is generated by a solenoid lens [304]. In contrast to PMQ's, the collimation can be obtained with a compact, single device.

In order to study the applicability of such a solenoid, simulations with CST Studio Suite 2008™ [287] have been performed. A flat disk of 100 µm diameter emitting (10 ± 1) MeV protons with 22° half angle divergence acts as a proton source. The solenoid was designed as a single coil with a 30 mm diameter aperture in the center, placed at 1 cm behind the source. The coil has 180 turns, a length of 8 cm and a soft iron core in order to enhance and homogenize the magnetic field. Figure 6.3 shows a schematic of the configuration. The current-carrying coil is the red cylinder. The laser beam (green triangle) is not part of the simulation and shown only for clarity. The coil is driven by a 10 kA current, leading to a magnetic field strength of 20 T. Since the ions are emitted under an angle from the source, there is a velocity component not being parallel to the homogeneous B-field in the coil. Hence the particles are being deflected according to the Lorentz force $\boldsymbol{F} = q \cdot \boldsymbol{v} \times \boldsymbol{B}$. Similar to the PMQs, the solenoid field works best for a discrete energy only. In the case here, most of the (10 ± 1) MeV protons could be collimated. The quality of the collimation is best seen in the transverse phase space plot.

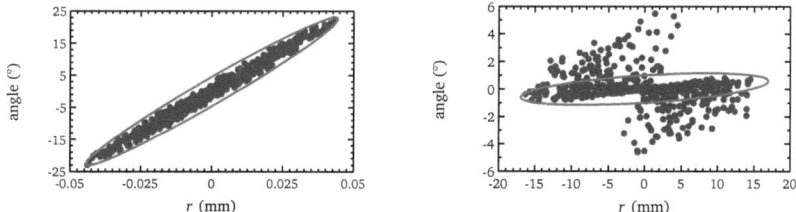

Fig. 6.4: Transverse phase space of sample particles in the magnetic field simulation. The left image shows the particle phase space just behind the source. The particles originate from a 100 µm diameter spot and are emitted with 22° HWHM angle. The right image shows the phase space several cm behind the coil. The beam is distributed in an area of about 30 mm diameter. The red ellipse depicts the region of most of the particles. The divergence angle is below 1°. The other particles with larger angles are the ones with energies deviating from the design value of 10 MeV.

Figure 6.4 shows the transverse phase space of sample particles in the magnetic field simulation. The left image shows the particle phase space just behind the source. The particles originate from a 100 μm diameter spot and are emitted with 22° HWHM angle. The red ellipse denotes area in the phase space. Figure 6.3 shows that behind the solenoid, the beam is mostly collimated. The right image in fig. 6.4 shows the phase space several cm behind the coil. The beam is distributed in an area of about 30 mm diameter as expected. The red ellipse depicts the region of most of the particles, with a divergence angle below 1° and particle energies close to 10 MeV. The other particles with larger angles are the ones with energies strongly deviating from the design value of 10 MeV, i.e., particles with either 9 MeV or 11 MeV. The other energies present in a real TNSA-beam will be either over-focused or not focused at all. A 3 cm diameter aperture could be placed behind the solenoid in order to monochromatize the beam.

Currently, a prototype of a solenoid coil is being developed by our group. Preliminary tests demonstrated a stable operation with a 13 T magnetic field strength, with the potential to further increase the field strength. Once the beam is collimated, it could be injected into a de-buncher section to further reduce the energy-spread and for post-acceleration or injection into a synchrotron [305–307].

6.2 Generation of high-energy density matter

The creation of extreme states of matter is important for the understanding of the physics covered in various research fields as high-pressure physics, applied material studies, planetary science, inertial fusion energy and all forms of plasma generation generated from solids. The primary difficulties in the study of these states of matter are, that the time scales or the changes are rapid ($\approx 1\,\mathrm{ps}$) while the matter is very dense and the temperatures are relatively low, on the order of a few eV/k_B. With these parameters, the plasma exhibits long- and short-range orders, that are due to the correlating effects of the ions and electrons. The state of matter is too dense and/or too cold to admit standard solutions used in plasma physics. Perturbative approaches using expansions in small parameters for the description of the plasma are no longer valid, providing a tremendous challenge for theoretical models. This region where condensed matter physics and plasma physics converge is the so-called *Warm Dense Matter* (WDM) regime [308].

WDM conditions can be generated in a number of ways, such as laser-generated shocks [7] or laser-generated x-rays [309, 310], intense ion beams from conventional accelerators [311] or laser-accelerated protons [29,30,78], just to name a few. Whereas lasers only interact with the surface of a sample, ions can penetrate deep into the material of interest thereby generating large samples of homogeneously heated matter. The short pulse duration of intense ion beams furthermore allows for the investigation of equation of states close to the solid state density, because of the material's inertia preventing the expansion of the sample within the interaction. In addition to that, the interaction of ions with matter dominantly is due to collisions and does not include a high temperature plasma corona as it is present in laser matter interaction. Due to these unique features of intense ion beam heating of matter, a collaboration for the investigation of High Energy-Density states of matter GEnerated by intense Heavy iOn and laser Beams (HEDgeHOB) at the future Facility for Anti-proton and Ion Research (FAIR) at Darmstadt, Germany [312] has been formed [313].

The generation of large homogenous samples of WDM is accompanied by the challenging task to diagnose this state of matter, as usual diagnostic techniques fail under these conditions. The material density results in a huge opacity and the relatively low temperature does not allow traditional spectroscopic methods to be applied. Moreover the sample size, deposited energy and lifetime of the matter state are strongly interrelated and dominated by the stagnation time of the atoms in the probe. Thus high spatial and temporal resolution is required to gain quantitative data in those experiments. Due to the high density of the sample, laser diagnostics cannot be used. The properties of matter could be determined by measuring the expansion after the heating [30] or by measuring the thermal radiation emitted by the sample [29]. However, even more interesting are the plasma parameters deep inside the sample, where the ion heating is most effective. An ideally suited diagnostic technique recently developed is *X-ray Thomson Scattering* [309, 314, 315]. The scattering of externally generated x-rays off electrons in the dense plasma has demonstrated excellent diagnostic quality. It is not only able to penetrate deep into the matter revealing the properties in the bulk material, but it also simultaneously results in the most wanted parameters temperature and density with highest precision. The challenge is the small cross section for the interaction which requires a powerful x-ray source, a background radiation level as well as high resolution spectrometers with high efficiency.

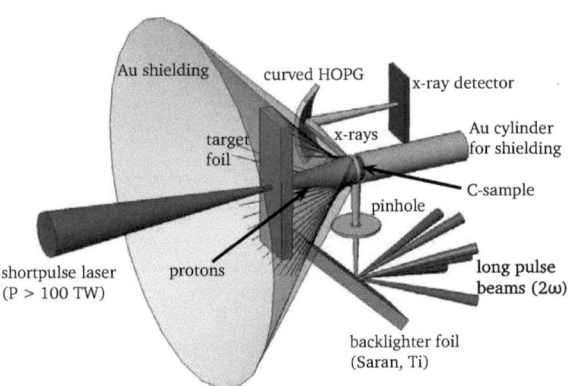

Fig. 6.5: Proposed experimental scheme to investigate the properties of laser-accelerated proton-heated matter by spectrally resolved x-ray Thomson scattering. See text for details. Image courtesy of K. Harres.

Figure 6.5 shows a proposed experimental scheme to investigate the transformation of solid, low-Z material into the WDM state. The experimental scheme requires a high-energy short-pulse laser and one or more long-pulse laser beams in the same experimental vacuum chamber. In recent years, more and more laser facilities have upgraded their laser systems for such kind of pump-probe experiments. A CPA laser beam above 100 TW power generates an intense proton beam from a thin target foil. The protons hit a solid density sample and heat it isochoric up to several eV/k_B temperature. The long-pulse beam(s) is (are) used to drive an intense x-ray source from a Ti or Saran (contains Cl) foil. The sample is probed by narrowband line-radiation from the Cl- or Ti-plasma. The scattered radiation is first spectrally dispersed by a highly efficient, highly-oriented pyrolytic graphite (HOPG) crystal spectrometer in von Hamos geometry before it is detected. Extensive gold shielding (partially shown above) is required to prevent parasitic signals in the scatter spectrometer. From the measured Doppler-broadened, Compton-downshifted signal the temperature and density can be inferred.

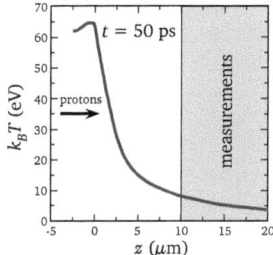 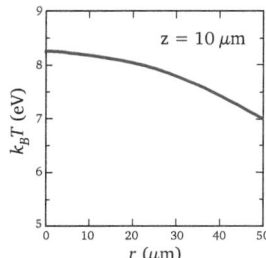

Fig. 6.6: Temperature distribution in the solid carbon sample, irradiated by TNSA-protons. See text for the proton beam parameters. The left image shows the temperature distribution along the longitudinal direction z at $t = 50$ ps after the first protons hit the target, The yellow-shaded area shows a region with constant temperature of about 8 eV, suitable for x-ray scattering measurements. In radial direction (see right-hand image) the temperature is homogeneous at around 8 eV, too. Image courtesy of An. Tauschwitz.

Two-dimensional hydrodynamic simulations of the laser-accelerated proton beam energy deposition in a solid carbon cylinder have been performed by An. Tauschwitz, in order to estimate the temperature by the proton heating. Details of the simulation are published in ref. [13]. For the estimate the proton beam parameters (spectrum, total number, energy dependent source size, energy dependent divergence) determined at LULI-100 TW have been used. The simulated proton beam had an exponential spectrum (eq. (3.13)) with $N_0 = 5 \times 10^{11}$ and $k_B T = 1.5$ MeV. The angle of beam spread decreases parabolic with energy. The decreasing source size with energy has been fit by a Gaussian. The distance between the proton producing foil and the carbon sample was chosen to 300 μm.

The temperature distribution in the solid carbon sample is shown in fig. 6.6. The left image shows the temperature distribution along the longitudinal direction z. 50 ps after the first protons hit the target, a plateau-like region (between 10 μm and 20 μm) of about 8 eV is formed in the yellow-shaded area. In radial direction (see right-hand image) the temperature is homogeneous at around 8 eV, too. Due to the impulsive heating by the protons the temperature stays constant for more than 300 ps . This is enough time for the backlighter beams to probe the sample.

Nevertheless, due to the close proximity of the proton source and the sample, artificial heating due to electrons and x-rays could occur [316]. Additionally, there are strong requirements on the shielding from parasitic radiation originating in the CPA-irradiated foil. Both drawbacks could be relaxed by using a strong solenoid magnet for transporting and focusing the protons from the source to the sample. The magnet would allow for a large distance between the proton-production foil and the sample, hence for clean experiments since possible pre-heating by high-energy photons and electrons would be significantly reduced. As an example [1], the HED state of matter that is exposed to 10^{11} protons with 15 MeV, focused to a small spot of 10 μm diameter was calculated with a semi-empirical equation-of-state model by N.A. Tahir [317]. 10 μm thin Al, Ni and Pb foils were chosen to sample matter from low to high nuclear charge. The energy of the protons is sufficient to penetrate the foil, and the energy deposition is very homogeneous since the Bragg-peak is outside the foil. The total specific energy deposition is 505.6 kJ/g for Al, 409.3 kJ/g for Ni and 277 kJ/g for Pb, respectively. The fast energy deposition by the protons leads to an isochoric heating, that transforms the former solid foil to liquid HED matter with temperatures around 20 eV/k_B ($\approx 2.5 \times 10^5$ K) and around 10 MBar pressure. This extreme state of matter could be found inside giant planets like Jupiter or Saturn, hence the heating of matter by laser-accelerated protons would allow to investigate states of matter that are of relevance for astrophysical questions.

Appendix A

Plasma Simulation Code - PSC

The intense laser-plasma interaction with relativistic intensities ($a_0 > 1$) generates plasma far away from equilibrium. The underlying physics is described by classical, nonlinear transport equations coupled with Maxwell fields. Quantum correlations can usually be neglected due to the very high temperatures involved. The extremely nonlinear nature of most of the phenomena involved in laser-matter interaction as e.g. creation of intense relativistic electrons as well as complex geometries make an analytical treatment extremely difficult. Therefore an approach based on computer simulations is most often the only possibility.

Luckily, in the last years the computing power has tremendously evolved. In addition to that, much faster program execution is obtained by the realization of massively parallel high-power computing platforms with multiple-core processors even for desktop computers. Together with the development of novel numerical schemes complex computer codes that include most aspects of super-intense laser-plasma interaction became possible. One of the most powerful codes is the *Plasma Simulation Code (PSC)*, developed by Hartmut Ruhl and Andreas Kemp [204, 318].

A.1 Governing equations

The plasma is considered consisting of ions and electrons, represented by distribution functions $f_k(\boldsymbol{x}, \boldsymbol{p}, t)$. The distribution functions f_k give the probability of finding particles of sort k in a given volume of the six-dimensional phase space. It is assumed that the electrons and ions in the plasma under consideration interact via electromagnetic forces only. Binary collisions will be neglected for simplicity. Hence, an appropriate description of the plasma is based on the relativistic Vlasov-equation:

$$\frac{\partial f_k}{\partial t} + \boldsymbol{v}_k \frac{\partial f_k}{\partial \boldsymbol{x}} + q_k \left(\boldsymbol{E} + \boldsymbol{v}_k \times \boldsymbol{B}\right) \frac{\partial f_k}{\partial \boldsymbol{p}_k} = 0 \,, \tag{A.1}$$

where the velocity $\boldsymbol{v}$ has to be expressed in terms of the momentum $\boldsymbol{p}$ as $\boldsymbol{v} = \boldsymbol{p}/m\left(1 + p^2/(m^2 c^2)\right)^{-1/2}$. The electromagnetic fields are obtained by solving the Maxwell equations

$$\partial_t \boldsymbol{E} = c^2 \nabla \times \boldsymbol{B} - \boldsymbol{j}/\varepsilon_0 \,, \tag{A.2}$$
$$\partial_t \boldsymbol{B} = -\nabla \times \boldsymbol{E} \,, \tag{A.3}$$
$$\partial_t \rho = -\nabla \cdot \boldsymbol{j} \,. \tag{A.4}$$

The system of equations is closed by integrating the zeroth and first moment of the distributions functions $f_k(\boldsymbol{x}, \boldsymbol{p}, t)$, to obtain the current density $\boldsymbol{j}$ and charge density ρ, respectively:

$$\rho = q_e \int f_e(\boldsymbol{x}, \boldsymbol{p}, t) \, \mathrm{d}^3 p_e + q_i \int f_i(\boldsymbol{x}, \boldsymbol{p}, t) \, \mathrm{d}^3 p_i \,, \tag{A.5}$$
$$\boldsymbol{j} = q_e \int \boldsymbol{v}_e f_e(\boldsymbol{x}, \boldsymbol{p}, t) \, \mathrm{d}^3 p_e + q_i \int \boldsymbol{v}_i f_i(\boldsymbol{x}, \boldsymbol{p}, t) \, \mathrm{d}^3 p_i \,. \tag{A.6}$$

A.2 Numerical implementation

The numerical approach used in the PSC to describe the nonlinear, kinetic nature of the interaction by solving the fully relativistic Vlasov equations combined with Maxwell's equations is a finite-element approach by the Particle-In-Cell (PIC) method. The code solves the Maxwell-Vlasov-Boltzmann equations with a method known as the Monte-Carlo Particle-In-Cell (MCPIC) approach, it includes binary particle collisions. The MCPIC method makes use of a mesh to represent the Maxwell fields and of finite elements or *quasi-particles*, each one representing several thousand real particles. The mesh is implemented as a three-dimensional array of arbitrary length, defined by the problem to solve and the computing power. The boundary conditions for the solutions are either periodic boundaries or reflecting ones. For the numerical solution the physical quantities from above are normalized with the help of the laser frequency ω_L, the speed of light c, the laser wavenumber $\tilde{\lambda}_L = c/\omega_L$, the laser's electric field amplitude and the laser's magnetic field amplitude, respectively. The governing equations are discretized and solved with second order accuracy in a finite-difference time-domain (FDTD) scheme. The energy-conservation is calculated by the Poynting-flux. Details of the implementation can be found in the code documentation [319] as well as in refs. [204, 318].

The source code is written in the Fortran90 language, since it is ideally suited and optimized for numerical calculation. The PSC features 3D domain decomposition, that means the computation arrays are divided among the number of CPUs on a distributed computing platform. Therefore the Message Passing Interface MPI is used, that is a message passing standard available on most workstation clusters and parallel supercomputers up to a few thousand compute nodes. For optimum load balancing the code has a sophisticated initialization routine that automatically divides and distributes the individual computing domains in a way that each CPU has approximately the same load. For safe operation, e.g. to avoid a complete restart in case a computing node crashes, the PSC checkpoints its core at predefined time intervals. At predefined time steps, the electromagnetic field data as well as the particle data are written on hard disk; a separate file is created for each node, for each particle sort and for the fields. The code allows to arbitrarily distribute these files over the nodes. For the data analysis, a separate post-processing routine is provided that collects the data and produces separate ASCII-files for each field and each density and current distribution in each direction x, y, z, as well as for each particle sort, respectively. These data can then be analyzed, e.g. with visualization software. The code and its analysis tools are written for a Unix operating system; all software is open source and freely available.

However, the PIC method has some drawbacks: it is extremely noisy [204] and too low resolution leads to artificial heating, the so-called *numerical heating* [133]. The way to overcome these problems are high spatial resolution and small time-steps which is on the cost of execution speed and has high requirements on computer hardware. The spatial resolution should ideally be less than the Debye length to correctly resolve electromagnetic interaction. For solid-density plasmas at, say $1\,\text{keV}/k_B$ temperature, the Debye length is $\lambda_D = 0.74\,\text{nm}$ only. Therefore the simulations performed during this thesis were restricted to two spatial dimensions. The electromagnetic fields are still being calculated in three dimensions, in a so-called 2½D configuration. Furthermore, the simulation runs were carried out without using the collisional module in PSC. The practical reason is, that in order to resolve the collisional processes correctly a huge number of cells and particles have to be used, tremendously slowing down the computation speed. On the other hand, since the plasmas investigated in the simulations have high temperatures, the probability for collisions is low. Sentoku *et al.* [36] have compared the collisional case and collisionless cases. The reference studied the lateral transport of fast electrons in a thin foil target. They found that there are two important lateral transport mechanisms, one is driven by the fast electrons' lateral diffusion and the other is due to the resistively driven fields. The fast electron distribution will become quite uniform in the target due to this fast lateral transport. It is shown that fast ion acceleration is mainly driven by these fast electron clouds, a strong evidence for using collisionless modeling of the fast ion acceleration.

A.3 Hard- and software installation

At the beginning of this thesis the plasma physics group at GSI did not have access to a computing cluster, and the existing computing cluster at GSI could not be set up for massively parallel running computations. Nevertheless, the application of the PSC to support the experiments on short-pulse laser-matter interaction seemed very promising.

Fig. A.1: Quad-Dualcore computing cluster.

Two computation servers were bought and installed as a cluster in the context of this work. Figure A.1 shows a photography of the cluster, installed in a rack with other machines at GSI. The two cluster nodes are the machines with the white, rectangular labels on the DVD drive.

Each cluster node is equipped with four dual-core AMD Opteron™ CPUs with 2 GHz clock rate. Each machine uses 32 GB RAM, resulting in 4 GB per CPU. The hard disk storage for each machine consists of a 160 GB hard disk for the operating system and four separate 500 GB hard disks in an RAID-5 array, leading to 1.5 TB for each server and summing up to 3 TB total hard disk storage. The RAID-5 arrays have hot-plug capability, meaning that malfunctioning hard disks can be exchanged without shutting down the server. The two nodes are prepared for a fast, 10 Gigabit Infiniband® network (http://www.infinibandta.org), however due to technical difficulties with the interfaces, both servers are connected via Gigabit fast ethernet connection in a private network. The Gigabit ethernet connection proved to be fast enough for the data transfer during PSC runs. Both machines are connected to the internal GSI network with a second network interface, allowing for remote access. The servers run a SuSE 10.2 Linux 64-bit operating system (http://www.opensuse.org). Each RAID-5 array is mounted as one large, XFS-formatted partition in the operating system. These two partitions are further cross-mounted via NFS on both machines, hence the PSC can access both large storage areas. Communication between the two nodes is provided by the mpich2 message passing interface software (http://www.mcs.anl.gov/research/projects/mpich2/) in the private network. The standard compiler is the GNU gfortran compiler (http://gcc.gnu.org/fortran/).

The system performance has been determined with the linpack HPL benchmark (http://top500.org/project/linpack), using the GotoBLAS linear algebra subroutine (http://www.tacc.utexas.edu/resources/software/gotoblasfaq.php). In summary, a single node reaches up to 2.88×10^{10} FLoating point Operations Per Second (28.8 GFLOPS). The two nodes together reach 55 GFLOPS, the deviation from twice the single-node performance is due to the Gigabit ethernet connection. The 55 GFLOPS total performance correspond to about 3.4 GFLOPS per CPU, a valued expected from data by the manufacturer [320].

Further software requirements of the PSC are for data analysis. The PSC stores its data in a Fortran binary format, that is additionally divided by the number of calculation CPUs allowing to distribute the data over different machines. Another Fortran program provided by the authors of PSC collects the data, sorts the individual fields and particle data and saves them as ASCII-files. The original version of PSC is shipped with scripts for the commercially available Interactive Data Language IDL (http://rsinc.com/idl/) software for data analysis and visualization. However, there are some drawbacks in using IDL. The license for IDL is expensive and the language is not very intuitive to learn. Due to these facts the author of this thesis has ported the IDL scripts to the Python scripting language (http://www.python.org). Python is a widely distributed, straightforward, open-source scripting language. Time-critical sections in the scripts, as e.g. the reading into memory and analysis of the huge particle data, are written in C and cross-linked to Python via SWIG (http://www.swig.org). Data processing is done with the Scipy package (http://www.scipy.org). Data visualization, either one-dimensional or two-dimensional plots, is performed with the Matplotlib extension to Python (http://matplotlib.sourceforge.net). The graphics can be saved in a variety of file formats, for convenience the PNG format is chosen for most plots.

An example shall now be used for an estimate on the tremendous computing resources needed: Assuming an interaction volume of $x \times y \times z = (5 \times 5 \times 10)\,\mu\text{m}^3$ in the simulation, about $6750^2 \times 13500 \approx 6 \times 10^{11}$ cells have to be defined to resolve the Debye length in the solid density plasma. The target is chosen to be only $1\,\mu\text{m}$ thin and $(5 \times 5)\,\mu\text{m}^2$ wide. It consists of protons and electrons only, with 1 quasi-particle per cell and per species. Thus, about 6.2×10^{10} quasi-particles will be generated. Each particle is represented in the PSC by 11 variables: position (x, y, z), momentum (p_x, p_y, p_z), charge q, mass m, its current cell number c_{n_i}, a unique particle label l_{n_i} and its particle weight w_{n_i}, respectively. Each variable is represented in Fortran double precision (8 bytes, meaning the code does 64-bit arithmetic). Therefore about $8 \times 11 \times 10^{-6}\,\text{MB} = 8.5 \times 10^{-5}\,\text{MB}$ are required per particle.

Furthermore, the code allocates 12 fields with Fortran double precision during the calculation: the electric fields (E_x, E_y, E_z), magnetic fields (B_x, B_y, B_z), current densities (j_x, j_y, j_z), charge densities (ρ_i, ρ_e) and the Poynting flux P. Additionally, 24 fields with single precision resolution are required for time-resolved and time-averaged fields that are written on hard disk. Hence $(24 \times 4 + 12 \times 8) \times 10^{-6}\,\text{MB} = 1.92 \times 10^{-4}\,\text{MB}$ per cell is required. In addition, about $250\,\text{MB}$ of memory are required for communication.

In summary, the example would require about 130 Terabyte computer memory, that just fits into the total memory of the current world's largest (not the fastest) supercomputer BlueGene/L at Lawrence Livermore National Laboratory, CA, USA [321]!

Bibliography

[1] **M. Schollmeier**, S. Becker, M. Geißel, K.A. Flippo, A. Blažević, S.A. Gaillard, D.C. Gautier, F. Grüner, K. Harres, M. Kimmel, F. Nürnberg, P. Rambo, U. Schramm, J. Schreiber, J. Schütrumpf, J. Schwarz, N.A. Tahir, B. Atherton, D. Habs, B.M. Hegelich, and M. Roth. *Controlled Transport and Focusing of Laser-Accelerated Protons with Miniature Magnetic Devices.* Phys. Rev. Lett. **101**, 055004 (2008).

[2] **M. Schollmeier**, K. Harres, F. Nürnberg, A. Blažević, P. Audebert, E. Brambrink, J.C. Fernández, K.A. Flippo, D.C. Gautier, M. Geißel, B.M. Hegelich, J. Schreiber, and M. Roth. *Laser beam-profile impression and target thickness impact on laser-accelerated protons.* Phys. Plasmas **15**, 053101 (2008).

[3] **Marius Schollmeier**, M. Roth, A. Blažević, E. Brambrink, J.A. Cobble, J.C. Fernandez, K.A. Flippo, D.C. Gautier, D. Habs, K. Harres, B.M. Hegelich, T. Heßling, D.H.H. Hoffmann, S. Letzring, F. Nürnberg, G. Schaumann, J. Schreiber, and K. Witte. *Laser ion acceleration with micro-grooved targets.* Nucl. Instrum. and Meth. A **577**, 186 (2007).

[4] K.A. Flippo, E. d'Humières, S.A. Gaillard, J. Rassuchine, D.C. Gautier, **M. Schollmeier**, F. Nürnberg, J.L. Kline, J. Adams, B. Albright, M. Bakeman, K. Harres, R.P. Johnson, G. Korgan, S. Letzring, S. Malekos, N. Renard-Legalloudec, Y. Sentoku, T. Shimada, M. Roth, T.E. Cowan, J.C. Fernández, and B.M. Hegelich. *Increased efficiency of short-pulse laser-generated proton beams from novel flat-top cone targets.* Phys. Plasmas **15**, 056709 (2008).

[5] K.A. Flippo, B.M. Hegelich, M.J. Schmitt, D.C. Gauthier, C.A. Meserole, G.L. Fisher, J.A. Cobble, R.A. Johnson, S.A. Letzring, J.C. Fernández, **M. Schollmeier**, and J. Schreiber. *Ablation cleaning techniques for high-power short-pulse laser-produced heavy ion targets.* Proceedings of SPIE **6261**, 62612I (2006).

[6] K.A. Flippo, B.M. Hegelich, M.J. Schmitt, C.A. Meserole, G.L. Fisher, D.C. Gautier, J.A. Cobble, R. Johnson, S. Letzring, J. Schreiber, **M. Schollmeier**, and J.C. Fernández. *Ultrashort-laser-produced heavy ion generation via target ablation cleaning.* Journal de Physique IV **133**, 1117 (2006).

[7] E. Garcia Saiz, G. Gregori, D.O. Gericke, J. Vorberger, B. Barbrel, R.J. Clarke, R.R. Freeman, S.H. Glenzer, F.Y. Khattak, M. Koenig, O.L. Landen, D. Neely, P. Neumayer, M.M. Notley, A. Pelka, D. Price, M. Roth, **M. Schollmeier**, C. Spindloe, R.L. Weber, L. van Woerkom, K. Wünsch, , and D. Riley. *Probing warm dense lithium by inelastic x-ray scattering.* Nature Phys. (2008). *accepted for publication.*

[8] K. Harres, **M. Schollmeier**, E. Brambrink, P. Audebert, A. Blažević, K. Flippo, D.C. Gautier, M. Geißel, B.M. Hegelich, F. Nürnberg, J. Schreiber, H. Wahl, and M. Roth. *Development and calibration of a Thomson parabola with microchannel plate for the detection of laser-accelerated MeV ions.* Rev. Sci. Instrum. **79**, 093306 (2008).

[9] F. Nürnberg, **M. Schollmeier**, K. Harres, M. Roth, A. Blažević, K. Witte, J. Schreiber, K. Flippo, B.M. Hegelich, D.C. Gautier, J.C. Fernández, E. Brambrink, P. Audebert, P. McKenna, D.C. Carroll, O. Lundh, K. Markey, D. Neely, and P. Norreys. *RCF imaging spectroscopy of laser-accelerated proton beams.* Rev. Sci. Instrum. (2008). in preparation.

[10] M. Roth, P. Audebert, A. Blazevic, E. Brambrink, J. Cobble, T.E. Cowan, J. Fernandez, J. Fuchs, M. Geissel, M. Hegelich, S. Karsch, H. Ruhl, **M. Schollmeier**, and R. Stephens. *Laser accelerated heavy particles – Tailoring of ion beams on a nano-scale.* Optics Communications **264**, 519 (2006).

[11] M. Roth, E. Brambrink, P. Audebert, M. Basko, A. Blazevic, R. Clarke, J. Cobble, T.E. Cowan, J. Fernandez, J. Fuchs, M. Hegelich, K. Ledingham, B.G. Logan, D. Neely, H. Ruhl, and **M. Schollmeier**. *Laser accelerated ions in ICF research prospects and experiments.* Plasma Phys. Control. Fusion **47**, 841 (2005).

[12] H. Ruhl, **M. Schollmeier**, and T. Cowan. *The characterization of laser-accelerated proton flows and their application to the measurement of electric and magnetic fields in the flow –* in preparation. (2008).

[13] An. Tauschwitz, E. Brambrink, J.A. Maruhn, M. Roth, **M. Schollmeier**, T. Schlegel, and A. Tauschwitz. *Laser-produced proton beams as a tool for equation-of-state studies of warm dense matter.* High Energy Density Physics **2**, 16 – 20 (2006).

[14] D. Strickland and G. Mourou. *Compression of amplified chirped optical pulses.* Optics Communications **56**, 219 (1985).

[15] J. Fuchs, P. Antici, E. d'Humières, E. Lefebvre, M. Borghesi, E. Brambrink, C.A. Cecchetti, M. Kaluza, V. Malka, M. Manclossi, S. Meyroneinc, P. Mora, J. Schreiber, T. Toncian, H. Pépin, and P. Audebert. *Laser-driven proton scaling laws and new paths towards energy increase.* Nature Phys. **2**, 48 (2006).

[16] L. Robson, P.T. Simpson, R.J. Clarke, K.W.D. Ledingham, F. Lindau, O. Lundh, T. McCanny, P. Mora, D. Neely, C.-G. Wahlström, M. Zepf, and P. McKenna. *Scaling of proton acceleration driven by petawatt-laser-plasma interactions.* Nature Phys. **3**, 58 (2007).

[17] J. Schreiber, F. Bell, F. Grüner, U. Schramm, M. Geissler, M. Schnürer, S. Ter-Avetisyan, B.M. Hegelich, J. Cobble, E. Brambrink, J. Fuchs, P. Audebert, and D. Habs. *Analytical Model for Ion Acceleration by High-Intensity Laser Pulses.* Phys. Rev. Lett. **97**, 045005 (2006).

[18] E. Clark, K. Krushelnick, J. Davies, M. Zepf, M. Tatarakis, F. Beg, A. Machacek, P. Norreys, M. Santala, I. Watts, and A. Dangor. *Measurements of energetic proton transport through magnetized plasma from intense laser interactions with solids.* Phys. Rev. Lett. **84**, 670–3 (2000).

[19] A. Maksimchuk, S. Gu, K. Flippo, D. Umstadter, and V.Yu. Bychenkov. *Forward Ion Acceleration in Thin Films Driven by a High-Intensity Laser.* Phys. Rev. Lett. **84**, 4108 (2000).

[20] E. Clark, K. Krushelnick, M. Zepf, F. Beg, M. Tatarakis, A. Machacek, M. Santala, I. Watts, P. Norreys, and A. Dangor. *Energetic heavy-Ion and proton generation from ultraintense laser-plasma interactions with solids.* Phys. Rev. Lett. **85**, 1654–7 (2000).

[21] R.A. Snavely, M.H. Key, S.P. Hatchett, T.E. Cowan, M. Roth, T.W. Phillips, M.A. Stoyer, E.A. Henry, T.C. Sangster, M.S. Singh, S.C. Wilks, A. MacKinnon, A. Offenberger, D.M. Pennington, K. Yasuike, A.B. Langdon, B.F. Lasinski, J. Johnson, M.D. Perry, and E.M. Campbell. *Intense high-energy proton beams from Petawatt-laser irradiation of solids.* Phys. Rev. Lett. **85**, 2945–8 (2000).

[22] S.P. Hatchett, C.G. Brown, T.E. Cowan, E.A. Henry, J.S. Johnson, M.H. Key, J.A. Koch, A.B. Langdon, B.F. Lasinski, R.W. Lee, A.J. MacKinnon, D.M. Pennington, M.D. Perry, T.W. Phillips, M. Roth, T.C. Sangster, M.S. Singh, R.A. Snavely, M.A. Stoyer, S.C. Wilks, and K. Yasuike. *Electron, photon, and ion beams from the relativistic interaction of Petawatt laser pulses with solid targets.* Phys. Plasmas **7**, 2076 (2000).

[23] K. Krushelnick, E.L. Clark, M. Zepf, J.R. Davies, F.N. Beg, A. Machacek, M.I.K Santala, M. Tatarakis, I. Watts, P.A. Norreys, and A.E. Dangor. *Energetic proton production from relativistic laser interaction with high density plasmas.* Phys. Plasmas **7**, 2055 (2000).

[24] T.E. Cowan, M. Roth, J. Johnson, C. Brown, M. Christl, W. Fountain, S. Hatchett, E.A. Henry, A.W. Hunt, M.H. Key, A. MacKinnon, T. Parnell, D.M. Pennington, M.D. Perry, T.W. Phillips, T.C. Sangster, M. Singh, R. Snavely, M. Stoyer, Y. Takahashi, S.C. Wilks, and K. Yasuike. *Intense electron and proton beams from PetaWatt laser-matter interactions.* Nucl. Instr. and Meth. A **455**, 130 (2000).

[25] M. Borghesi, A. Schiavi, D.H. Campbell, M.G. Haines, O. Willi, A.J. Mackinnon, L.A. Gizzi, M. Galimberti, R.J. Clarke, and H. Ruhl. *Proton imaging: a diagnostic for inertial confinement fusion/fast ignitor studies.* Plasma Phys. Control. Fusion **43**, 267 (2001).

[26] L. Romagnani, J. Fuchs, M. Borghesi, P. Antici, P. Audebert, F. Ceccherini, T. Cowan, T. Grismayer, S. Kar, A. Macchi, P. Mora, G. Pretzler, A. Schiavi, T. Toncian, and O. Willi. *Dynamics of electric fields driving the laser acceleration of multi-MeV protons.* Phys. Rev. Lett. **95**, 195001 (2005).

[27] A.J. Mackinnon, P.K. Patel, M. Borghesi, R.C. Clarke, R.R. Freeman, H. Habara, S.P. Hatchett, D. Hey, D.G. Hicks, S. Kar, M.H. Key, J.A. King, K. Lancaster, D. Neely, A. Nikkro, P.A. Norreys, M.M. Notley, T.W. Phillips, L. Romagnani, R.A. Snavely, R.B. Stephens, and R.P.J. Town. *Proton Radiography of a Laser-Driven Implosion.* Phys. Rev. Lett. **97**, 045001 (2006).

[28] L. Romagnani, S.V. Bulanov, M. Borghesi, P. Audebert, J.C. Gauthier, K. Löwenbrück, A.J. Mackinnon, P. Patel, G. Pretzler, T. Toncian, and O. Willi. *Observation of Collisionless Shocks in Laser-Plasma Experiments.* Phys. Rev. Lett. **101**, 025004 (2008).

[29] P.K. Patel, A.J. Mackinnon, M.H. Key, T.E. Cowan, M.E. Foord, M. Allen, D.F. Price, H. Ruhl, P.T. Springer, and R. Stephens. *Isochoric heating of solid-density matter with an ultrafast proton beam*. Phys. Rev. Lett. **91**, 125004 (2003).

[30] G.M. Dyer, A.C. Bernstein, B.I. Cho, J. Osterholz, W. Grigsby, A. Dalton, R. Shepherd, Y. Ping, H. Chen, K. Widmann, and T. Ditmire. *Equation-of-State Measurement of Dense Plasmas Heated With Fast Protons*. Phys. Rev. Lett. **101**, 015002 (2008).

[31] A. Pukhov. *Three-Dimensional Simulations of Ion Acceleration from a Foil Irradiated by a Short-Pulse Laser*. Phys. Rev. Lett. **86**, 3562 (2001).

[32] K. Krushelnick, E.L. Clark, R. Allott, F.N. Beg, C.N. Danson, A. Machacek, V. Malka, Z. Najmudin, D. Neely, P.A. Norreys, M.R. Salvati, M.I.K. Santala, M. Tatarakis, I. Watts, M. Zepf, and A.E. Dangor. *Ultrahigh-intensity laser-produced plasmas as a compact heavy ion injection source*. IEEE Transactions on Plasma Science **28**, 1184 (2000).

[33] D. Habs, G. Pretzler, A. Pukhov, and J. Meyer-Ter-Vehn. *Laser acceleration of electrons and ions and intense secondary particle generation*. Progress in Particle and Nuclear Physics **46**, 375 (2001).

[34] I. Spencer, K.W.D. Ledingham, R.P. Singhal, T. McCanny, P. McKenna, E.L. Clark, K. Krushelnick, M. Zepf, F.N. Beg, M. Tatarakis, A.E. Dangor, P.A. Norreys, R.J. Clarke, R.M. Allott, and I.N. Ross. *Laser generation of proton beams for the production of short-lived positron emitting radioisotopes*. Nucl. Instrum. and Meth. B **183**, 449 (2001).

[35] S. Karsch, S. Düsterer, H. Schwoerer, F. Ewald, D. Habs, M. Hegelich, G. Pretzler, A. Pukhov, K. Witte, and R. Sauerbrey. *High-Intensity Laser Induced Ion Acceleration from Heavy-Water Droplets*. Phys. Rev. Lett. **91**, 015001 (2003).

[36] Y. Sentoku, A.J. Kemp, R. Presura, M.S. Bakeman, and T.E. Cowan. *Isochoric heating in heterogeneous solid targets with ultrashort laser pulses*. Phys. Plasmas **14**, 122701 (2007).

[37] M.I.K Santala, M. Zepf, F.N. Beg, E.L. Clark, A.E. Dangor, K. Krushelnick, M. Tatarakis, I. Watts, K.W.D. Ledingham, T. McCanny, I. Spencer, A.C. Machacek, R. Allott, R.J. Clarke, and P.A. Norreys. *Production of radioactive nuclides by energetic protons generated from intense laser-plasma interactions*. Appl. Phys. Lett. **78**, 19 (2001).

[38] K. Nemoto, A. Maksimchuk, S. Banerjee, K. Flippo, G. Mourou, D. Umstadter, and V.Yu. Bychenkov. *Laser-triggered ion acceleration and table top isotope production*. Appl. Phys. Lett. **78**, 595 (2001).

[39] K.W.D. Ledingham, P. McKenna, T. McCanny, S. Shimizu, J.M. Yang, L. Robson, J. Zweit, J.M. Gillies, J. Bailey, G.N. Chimon, R.J. Clarke, D. Neely, P.A. Norreys, J.L. Collier, R.P. Singhal, M.S. Wei, S.P.D. Mangles, P. Nilson, K. Krushelnick, and M. Zepf. *High power laser production of short-lived isotopes for positron emission tomography*. J. Phys. D: Appl. Phys. **37**, 2341 (2004).

[40] S. Fritzler, V. Malka, G. Grillon, J.P. Rousseau, F. Burgy, E. Lefebvre, E. d'Humières, P. McKenna, and K.W.D. Ledingham. *Proton beams generated with high-intensity lasers: Applications to medical isotope production*. Appl. Phys. Lett. **83**, 3039 (2003).

[41] P. McKenna, K.W.D. Ledingham, T. McCanny, R.P. Singhal, I. Spencer, E.L. Clark, F.N. Beg, K. Krushelnick, M.S. Wei, J. Galy, J. Magill, R.J. Clarke, K.L. Lancaster, P.A. Norreys, K. Spohr, and R. Chapman. *Effect of target heating on ion-induced reactions in high-intensity laser-plasma interactions.* Appl. Phys. Lett. **83**, 2763 (2003).

[42] Gregory Lapicki. *Scaling of analytical cross sections for K-shell ionization by nonrelativistic protons to cross sections by protons at relativistic velocities.* J. Phys. B: At. Mol. Opt. Phys. **41**, 115201 (2008).

[43] M. Roth, T.E. Cowan, M.H. Key, S.P. Hatchett, C. Brown, W. Fountain, J. Johnson, D.M. Pennington, R.A. Snavely, S.C. Wilks, K. Yasuike, H. Ruhl, F. Pegoraro, S.V. Bulanov, E.M. Campbell, M.D. Perry, and H. Powell. *Fast ignition by intense laser-accelerated proton beams.* Phys. Rev. Lett. **86**, 436 (2001).

[44] V.Yu Bychenkov, W. Rozmus, A. Maksimchuk, D. Umstadter, and C.E. Capjack. *Fast Ignitor Concept with Light Ions.* Plasma Physics Reports **27**, 1017 (2001).

[45] E. Fourkal, B. Shahine, M. Ding, J.S. Li, T. Tajima, and C.M. Ma. *Particle in cell simulation of laser-accelerated proton beams for radiation therapy.* Med. Phys. **29**, 2788–98 (2002).

[46] S.V. Bulanov and V.S. Khoroshkov. *Feasibility of using laser ion accelerators in proton therapy.* Plasma Physics Reports **28**, 453 – 456 (2002).

[47] S.S. Bulanov, A. Brantov, V.Yu Bychenkov, V. Chvykov, G. Kalinchenko, T. Matsuoka, P. Rousseau, S. Reed, V. Yanovsky, K. Krushelnick, D.W. Litzenberg, and A.y Maksimchuk. *Accelerating protons to therapeutic energies with ultraintense, ultraclean, and ultrashort laser pulses.* Med. Phys. **35**, 1770–6 (2008).

[48] M. Murakami, Y. Hishikawa, S. Miyajima, Y. Okazaki, K.L. Sutherland, M. Abe, S.V. Bulanov, H. Daido, T.Zh. Esirkepov, J. Koga, M. Yamagiwa, and T. Tajima. *Radiotherapy using a laser proton accelerator.* eprint arXiv **0804**, 3826 (2008).

[49] M. Hegelich, S. Karsch, G. Pretzler, D. Habs, K. Witte, W. Guenther, M. Allen, A. Blažević, J. Fuchs, J.C. Gauthier, M. Geissel, P. Audebert, T. Cowan, and M. Roth. *MeV ion jets from short-pulse-laser interaction with thin foils.* Phys. Rev. Lett. **89**, 085002 (2002).

[50] J.C. Fernández, B.M. Hegelich, J.A. Cobble, K.A. Flippo, S.A. Letzring, R.P. Johnson, D.C. Gautier, T. Shimada, G.A. Kyrala, Y. Wang, C.J. Wetteland, and J. Schreiber. *Laser-ablation treatment of short-pulse laser targets: Toward an experimental program on energetic-ion interactions with dense plasmas.* Laser Part. Beams **23**, 267 – 73 (2005).

[51] S.J. Gitomer, R.D. Jones, F. Begay, A.W. Ehler, J.F. Kephart, and R. Kristal. *Fast ions and hot electrons in the laser-plasma interaction.* Phys. Fluids **29**, 2679 (1986).

[52] S.C. Wilks, A.B. Langdon, T.E. Cowan, M. Roth, M. Singh, S. Hatchett, M.H. Key, D. Pennington, A. MacKinnon, and R.A. Snavely. *Energetic proton generation in ultra-intense laser-solid interactions.* Phys. Plasmas **8**, 542 (2001).

[53] A.J. Mackinnon, M. Borghesi, S. Hatchett, M.H. Key, P.K. Patel, H. Campbell, A. Schiavi, R. Snavely, S.C. Wilks, and O. Willi. *Effect of plasma scale length on multi-MeV proton production by intense laser pulses.* Phys. Rev. Lett. **86**, 1769 (2001).

[54] M. Roth, A. Blazevic, M. Geissel, T. Schlegel, T.E. Cowan, M. Allen, J.-C. Gauthier, P. Audebert, J. Fuchs, J. Meyer-Ter-Vehn, M. Hegelich, S. Karsch, and A. Pukhov. *Energetic ions generated by laser pulses: A detailed study on target properties.* Physical Review Special Topics - Accelerators and Beams **5**, 061301 (2002).

[55] M. Allen, P.K. Patel, A. Mackinnon, D. Price, S. Wilks, and E. Morse. *Direct experimental evidence of back-surface ion acceleration from laser-irradiated gold foils.* Phys. Rev. Lett. **93**, 265004 (2004).

[56] Y. Murakami, Y. Kitagawa, Y. Sentoku, M. Mori, R. Kodama, K.A. Tanaka, K. Mima, and T. Yamanaka. *Observation of proton rear emission and possible gigagauss scale magnetic fields from ultra-intense laser illuminated plastic target.* Phys. Plasmas **8**, 4138 (2001).

[57] E. Brambrink, J. Schreiber, T. Schlegel, P. Audebert, J. Cobble, J. Fuchs, M. Hegelich, and M. Roth. *Transverse characteristics of short-pulse laser-produced ion beams: a study of the acceleration dynamics.* Phys. Rev. Lett. **96**, 154801 (2006).

[58] M. Tatarakis, J.R. Davies, P. Lee, P.A. Norreys, N.G. Kassapakis, F.N. Beg, A.R. Bell, M.G. Haines, and A.E. Dangor. *Plasma formation on the front and rear of plastic targets due to high-intensity laser-generated fast electrons.* Phys. Rev. Lett. **81**, 999 – 1002 (1998).

[59] M. Kaluza, J. Schreiber, M.I. Santala, G.D. Tsakiris, K. Eidmann, J. Meyer-Ter-Vehn, and K.J. Witte. *Influence of the Laser Prepulse on Proton Acceleration in Thin-Foil Experiments.* Phys. Rev. Lett. **93**, 045003 (2004).

[60] J. Fuchs, Y. Sentoku, S. Karsch, J. Cobble, P. Audebert, A. Kemp, A. Nikroo, P. Antici, E. Brambrink, A. Blažević, E.M. Campbell, J.C. Fernández, J.-C. Gauthier, M. Geissel, M. Hegelich, H. Pépin, H. Popescu, N. Renard-LeGalloudec, M. Roth, J. Schreiber, R. Stephens, and T.E. Cowan. *Comparison of laser ion acceleration from the front and rear surfaces of thin foils.* Phys. Rev. Lett. **94**, 045004 (2005).

[61] J. Fuchs, Y. Sentoku, E. d'Humières, T.E. Cowan, J. Cobble, P. Audebert, A. Kemp, A. Nikroo, P. Antici, E. Brambrink, A. Blažević, E.M. Campbell, J.C. Fernández, J.-C. Gauthier, M. Geissel, M. Hegelich, S. Karsch, H. Popescu, N. Renard-LeGalloudec, M. Roth, J. Schreiber, R. Stephens, and H. Pépin. *Comparative spectra and efficiencies of ions laser-accelerated forward from the front and rear surfaces of thin solid foils.* Phys. Plasmas **14**, 053105 (2007).

[62] Y. Sentoku, T.E. Cowan, A. Kemp, and H. Ruhl. *High energy proton acceleration in interaction of short laser pulse with dense plasma target.* Phys. Plasmas **10**, 2009 (2003).

[63] P. McKenna, D.C. Carroll, R.J. Clarke, R.G. Evans, K.W.D. Ledingham, F. Lindau, O. Lundh, T. McCanny, D. Neely, A.P.L. Robinson, L. Robson, P.T. Simpson, C.-G. Wahlström, and M. Zepf. *Lateral electron transport in high-intensity laser-irradiated foils diagnosed by ion emission.* Phys. Rev. Lett. **98**, 145001 (2007).

[64] F.N. Beg, M.S. Wei, E.L. Clark, A.E. Dangor, R.G. Evans, P. Gibbon, A. Gopal, K.L. Lancaster, K.W.D. Ledingham, P. McKenna, P.A. Norreys, M. Tatarakis, M. Zepf, and K. Krushelnick. *Return current and proton emission from short pulse laser interactions with wire targets.* Phys. Plasmas **11**, 2806 (2004).

[65] A.J. Mackinnon, Y. Sentoku, P.K. Patel, D.W. Price, S. Hatchett, M.H. Key, C. Andersen, R. Snavely, and R.R. Freeman. *Enhancement of Proton Acceleration by Hot-Electron Recirculation in Thin Foils Irradiated by Ultraintense Laser Pulses.* Phys. Rev. Lett. **88**, 215006 (2002).

[66] M.A. Stoyer, T.C. Sangster, E.A. Henry, M.D. Cable, T.E. Cowan, S.P. Hatchett, M.H. Key, M.J. Moran, D.M. Pennington, M.D. Perry, T.W. Phillips, M.S. Singh, R.A. Snavely, M. Tabak, and S.C. Wilks. *Nuclear diagnostics for petawatt experiments (invited).* Rev. Sci. Instrum. **72**, 767 (2001).

[67] P. Mora. *Plasma expansion into a vacuum.* Phys. Rev. Lett. **90**, 185002 (2003).

[68] P. Mora. *Collisionless expansion of a Gaussian plasma into a vacuum.* Phys. Plasmas **12**, 112102 (2005).

[69] P. Mora. *Thin-foil expansion into a vacuum.* Phys. Rev. E **72**, 056401 (2005).

[70] T. Grismayer and P. Mora. *Influence of a finite initial ion density gradient on plasma expansion into a vacuum.* Phys. Plasmas **13**, 032103 (2006).

[71] T. Grismayer, P. Mora, J.C. Adam, and A. Héron. *Electron kinetic effects in plasma expansion and ion acceleration.* Phys. Rev. E **77**, 066407 (2008).

[72] T.E. Cowan, J. Fuchs, H. Ruhl, A. Kemp, P. Audebert, M. Roth, R. Stephens, I. Barton, A. Blažević, E. Brambrink, J. Cobble, J. Fernandez, J.-C. Gauthier, M. Geissel, M. Hegelich, J. Kaae, S. Karsch, G.P. Le Sage, S. Letzring, M. Manclossi, S. Meyroneinc, A. Newkirk, H. Pepin, and N. Renard-LeGalloudec. *Ultralow emittance, multi-MeV proton beams from a laser virtual-cathode plasma accelerator.* Phys. Rev. Lett. **92**, 204801 (2004).

[73] T.E. Cowan, J. Fuchs, H. Ruhl, Y. Sentoku, A. Kemp, P. Audebert, M. Roth, R. Stephens, I. Barton, A. Blažević, E. Brambrink, J. Cobble, J.C. Fernandez, J.-C. Gauthier, M. Geissel, M. Hegelich, J. Kaae, S. Karsch, G.P. Le Sage, S. Letzring, M. Manclossi, S. Meyroneinc, A. Newkirk, H. Pepin, and N. Renard-LeGalloudec. *Ultra-low emittance, high current proton beams produced with a laser-virtual cathode sheath accelerator.* Nucl. Instr. and Meth. A **544**, 277 – 284 (2005).

[74] M. Borghesi, A.J. Mackinnon, D.H. Campbell, D.G. Hicks, S. Kar, P.K. Patel, D. Price, L. Romagnani, A. Schiavi, and O. Willi. *Multi-MeV proton source investigations in ultraintense laser-foil interactions.* Phys. Rev. Lett. **92**, 055003 (2004).

[75] J. Schreiber, M. Kaluza, F. Grüner, U. Schramm, B.M. Hegelich, J. Cobble, M. Geissler, E. Brambrink, J. Fuchs, P. Audebert, D. Habs, and K. Witte. *Source-size measurements and charge distributions of ions accelerated from thin foils irradiated by high-intensity laser pulses.* Appl. Phys. B **79**, 1041 (2004).

[76] J. Fuchs, T.E. Cowan, P. Audebert, H. Ruhl, L. Gremillet, A. Kemp, M. Allen, A. Blažević, J.-C. Gauthier, M. Geissel, M. Hegelich, S. Karsch, P. Parks, M. Roth, Y. Sentoku, R. Stephens, and E.M. Campbell. *Spatial uniformity of laser-accelerated ultrahigh-current MeV electron propagation in metals and insulators.* Phys. Rev. Lett. **91**, 255002 (2003).

[77] Y. Oishi, T. Nayuki, T. Fujii, Y. Takizawa, X. Wang, T. Yamazaki, K. Nemoto, T. Sekiya, K. Horioka, and A.A. Andreev. *Measurement of source profile of proton beams generated by ultraintense laser pulses using a Thomson mass spectrometer.* J. Appl. Phys. **97**, 4906 (2005).

[78] P. Antici, J. Fuchs, S. Atzeni, A. Benuzzi, E. Brambrink, M. Esposito, M. Koenig, A. Ravasio, J. Schreiber, A. Schiavi, and P. Audebert. *Isochoric heating of matter by laser-accelerated high-energy protons.* J. Phys. IV France **133**, 1077–79 (2006).

[79] M.H. Key. *Status of and prospects for the fast ignition inertial fusion concept.* Phys. Plasmas **14**, 055502 (2007).

[80] R.A. Snavely, B. Zhang, K. Akli, Z. Chen, R.R. Freeman, P. Gu, S.P. Hatchett, D. Hey, J. Hill, M.H. Key, Y. Izawa, J. King, Y. Kitagawa, R. Kodama, A.B. Langdon, B.F. Lasinski, A. Lei, A.J. Mackinnon, P. Patel, R. Stephens, M. Tampo, K.A. Tanaka, R. Town, Y. Toyama, T. Tsutsumi, S.C. Wilks, T. Yabuuchi, and J. Zheng. *Laser generated proton beam focusing and high temperature isochoric heating of solid matter.* Phys. Plasmas **14**, 092703 (2007).

[81] T. Toncian, M. Borghesi, J. Fuchs, E. d'Humières, P. Antici, P. Audebert, E. Brambrink, C.A. Cecchetti, A. Pipahl, L. Romagnani, and O. Willi. *Ultrafast laser-driven microlens to focus and energy-select mega-electron volt protons.* Science **312**, 410–3 (2006).

[82] S. Kar, K. Markey, P.T. Simpson, C. Bellei, J.S. Green, S.R. Nagel, S. Kneip, D.C. Carroll, B. Dromey, L. Willingale, E.L. Clark, P. McKenna, Z. Najmudin, K. Krushelnick, P. Norreys, R.J. Clarke, D. Neely, M. Borghesi, and M. Zepf. *Dynamic control of laser-produced proton beams.* Phys. Rev. Lett. **100**, 105004 (2008).

[83] T.E. Cowan, J. Fuchs, H. Ruhl, Y. Sentoku, A. Kemp, P. Audebert, M. Roth, R. Stephens, I. Barton, A. Blažević, E. Brambrink, J. Cobble, J.C. Fernández, J.-C. Gauthier, M. Geissel, M. Hegelich, J. Kaae, S. Karsch, G.P. Le Sage, S. Letzring, M. Manclossi, S. Meyroneinc, A. Newkirk, H. Pépin, and N. Renard-LeGalloudec. *Ultra-low emittance, high current proton beams produced with a laser-virtual cathode sheath accelerator.* Nucl. Instrum. and Meth. A **544**, 277 (2005).

[84] H. Eickhoff, Th. Haberer, B. Schlitt, and U. Weinrich. *HICAT - The german hospital-based light ion cancer therapy project.* Proceedings of EPAC, 290 (2004).

[85] T. Winkelmann, R. Cee, Th. Haberer, B. Naas, A. Peters, S. Scheloske, P. Spädtke, and K. Tinschert. *Electron cyclotron resonance ion source experience at the Heidelberg Ion Beam Therapy Center.* Rev. Sci. Instrum. **79**, 02A331 (2008).

[86] A.J. Kemp, J. Fuchs, Y. Sentoku, V. Sotnikov, M. Bakeman, P. Antici, and T.E. Cowan. *Emittance growth mechanisms for laser-accelerated proton beams.* Phys. Rev. E **75**, 056401 (2007).

[87] H. Ruhl, T. Cowan, and J. Fuchs. *The generation of micro-fiducials in laser-accelerated proton flows, their imaging property of surface structures and application for the characterization of the flow.* Phys. Plasmas **11**, L17 (2004).

[88] H. Ruhl, T. Cowan, and F. Pegoraro. *The generation of images of surface structures by laser-accelerated protons.* Laser Part. Beams **24**, 181 – 4 (2006).

[89] M.S. Wei, J.R. Davies, E.L. Clark, F.N. Beg, A. Gopal, M. Tatarakis, L. Willingale, P. Nilson, A.E. Dangor, P.A. Norreys, M. Zepf, and K. Krushelnick. *Reduction of proton acceleration in high-intensity laser interaction with solid two-layer targets.* Phys. Plasmas **13**, 123101 (2006).

[90] M. Allen, Y. Sentoku, P. Audebert, A. Blažević, T. Cowan, J. Fuchs, J.C. Gauthier, M. Geissel, M. Hegelich, S. Karsch, E. Morse, P.K. Patel, and M. Roth. *Proton spectra from ultraintense laser-plasma interaction with thin foils: Experiments, theory, and simulation.* Phys. Plasmas **10**, 3283 (2003).

[91] V.T. Tikhonchuk, A.A. Andreev, S.G. Bochkarev, and V.Yu. Bychenkov. *Ion acceleration in short-laser-pulse interaction with solid foils.* Plasma Phys. Control. Fusion **47**, 869 (2005).

[92] A.P. Robinson, A.R. Bell, and R.J. Kingham. *Effect of target composition on proton energy spectra in ultraintense laser-solid interactions.* Phys. Rev. Lett. **96**, 035005 (2006).

[93] M. Schnürer, S. Ter-Avetisyan, P.V. Nickles, and A.A. Andreev. *Influence of target system on the charge state, number, and spectral shape of ion beams accelerated by femtosecond high-intensity laser pulses.* Phys. Plasmas **14**, 033101 (2007).

[94] S. Ter-Avetisyan, M. Schnürer, P.V. Nickles, M. Kalashnikov, E. Risse, T. Sokollik, W. Sandner, A. Andreev, and V. Tikhonchuk. *Quasimonoenergetic deuteron bursts produced by ultraintense laser pulses.* Phys. Rev. Lett. **96**, 145006 (2006).

[95] A.J. Kemp and H. Ruhl. *Multispecies ion acceleration off laser-irradiated water droplets.* Phys. Plasmas **12**, 033105 (2005).

[96] S. Busch, O. Shiryaev, S. Ter-Avetisyan, M. Schnürer, P.V. Nickles, and W. Sandner. *Shape of ion energy spectra in ultra-short and intense laser-matter interaction.* Applied Physics B: Lasers and Optics **78**, 911 (2004).

[97] T.Zh. Esirkepov, S.V. Bulanov, K. Nishihara, T. Tajima, F. Pegoraro, V.S. Khoroshkov, K. Mima, H. Daido, Y. Kato, Y. Kitagawa, K. Nagai, and S. Sakabe. *Proposed double-layer target for the generation of high-quality laser-accelerated ion beams.* Phys. Rev. Lett. **89**, 175003 (2002).

[98] T. Morita, T.Zh. Esirkepov, S.V. Bulanov, J. Koga, and M. Yamagiwa. *Tunable high-energy ion source via oblique laser pulse incident on a double-layer target.* Phys. Rev. Lett. **100**, 145001 (2008).

[99] H. Schwoerer, S. Pfotenhauer, O. Jäckel, K.-U. Amthor, B. Liesfeld, W. Ziegler, R. Sauerbrey, K.W.D. Ledingham, and T. Esirkepov. *Laser-plasma acceleration of quasi-monoenergetic protons from microstructured targets.* Nature **439**, 445–8 (2006).

[100] B.M. Hegelich, B.J. Albright, J. Cobble, K. Flippo, S. Letzring, M. Paffett, H. Ruhl, J. Schreiber, R.K. Schulze, and J.C. Fernández. *Laser acceleration of quasi-monoenergetic MeV ion beams.* Nature **439**, 441–4 (2006).

[101] B.J. Albright, L. Yin, B.M. Hegelich, K.J. Bowers, T.J.T. Kwan, and J.C. Fernández. *Theory of laser acceleration of light-ion beams from interaction of ultrahigh-intensity lasers with layered targets.* Phys. Rev. Lett. **97**, 115002 (2006).

[102] A.P.L. Robinson and P. Gibbon. *Production of proton beams with narrow-band energy spectra from laser-irradiated ultrathin foils.* Phys. Rev. E **75**, 015401(R) (2007).

[103] S.M. Pfotenhauer, O. Jäckel, A. Sachtleben, J. Polz, W. Ziegler, H.-P. Schlenvoigt, K.-U. Amthor, M.C. Kaluza, K.W.D. Ledingham, R. Sauerbrey, P. Gibbon, A.P.L. Robinson, and H. Schwoerer. *Spectral shaping of laser generated proton beams.* New J. Phys. **10**, 3034 (2008).

[104] B.M. Hegelich, B. Albright, P. Audebert, A. Blažević, E. Brambrink, J. Cobble, T. Cowan, J. Fuchs, J.C. Gauthier, C. Gautier, M. Geissel, D. Habs, R. Johnson, S. Karsch, A. Kemp, S. Letzring, M. Roth, U. Schramm, J. Schreiber, K.J. Witte, and J.C. Fernández. *Spectral properties of laser-accelerated mid-Z MeV/u ion beams.* Phys. Plasmas **12**, 056314 (2005).

[105] M. Roth, T.E. Cowan, C. Brown, M. Christl, W. Fountain, S. Hatchett, J. Johnson, M.H. Key, D.M. Pennington, M.D. Perry, T.W. Phillips, T.C. Sangster, M. Singh, R. Snavely, M. Stoyer, Y. Takahashi, S.C. Wilks, and K. Yasuike. *Intense ion beams accelerated by Petawatt-class Lasers.* Nucl. Instrum. and Meth. A **464**, 201 (2001).

[106] D.C. Carroll, P. McKenna, O. Lundh, F. Lindau, C.-G. Wahlström, S. Bandyopadhyay, D. Pepler, D. Neely, S. Kar, P.T. Simpson, K. Markey, M. Zepf, C. Bellei, R.G. Evans, R. Redaelli, D. Batani, M.H. Xu, and Y.T. Li. *Active manipulation of the spatial energy distribution of laser-accelerated proton beams.* Phys. Rev. E **76**, 065401(R) (2007).

[107] O. Lundh, F. Lindau, A. Persson, C.-G. Wahlström, P. McKenna, and D. Batani. *Influence of shock waves on laser-driven proton acceleration.* Phys. Rev. E **76**, 026404 (2007).

[108] M.H. Xu, Y.T. Li, X.H. Yuan, Q.Z. Yu, S.J. Wang, W. Zhao, X.L. Wen, G.C. Wang, C.Y. Jiao, Y.L. He, S.G. Zhang, X.X. Wang, W.Z. Huang, Y.Q. Gu, and J. Zhang. *Effects of shock waves on spatial distribution of proton beams in ultrashort laser-foil interactions.* Phys. Plasmas **13**, 104507 (2006).

[109] J. Fuchs, C.A. Cecchetti, M. Borghesi, T. Grismayer, E. d'Humières, P. Antici, S. Atzeni, P. Mora, A. Pipahl, L. Romagnani, A. Schiavi, Y. Sentoku, T. Toncian, P. Audebert, and O. Willi. *Laser-foil acceleration of high-energy protons in small-scale plasma gradients.* Phys. Rev. Lett. **99**, 015002 (2007).

[110] A.A. Andreev, S. Sonobe, S. Kawata, S. Miyazaki, K. Sakai, K. Miyauchi, T. Kikuchi, K. Platonov, and K. Nemoto. *Effect of a laser prepulse on fast ion generation in the interaction of ultra-short intense laser pulses with a limited-mass foil target.* Plasma Phys. Control. Fusion **48**, 1605 (2006).

[111] K. Matsukado, T. Esirkepov, K. Kinoshita, H. Daido, T. Utsumi, Z. Li, A. Fukumi, Y. Hayashi, S. Orimo, M. Nishiuchi, S.V. Bulanov, T. Tajima, A. Noda, Y. Iwashita, T. Shirai, T. Takeuchi, S. Nakamura, A. Yamazaki, M. Ikegami, T. Mihara, A. Morita, M. Uesaka, K. Yoshii, T. Watanabe, T. Hosokai, A. Zhidkov, A. Ogata, Y. Wada, and T. Kubota. *Energetic protons from a few-micron metallic foil evaporated by an intense laser pulse.* Phys. Rev. Lett. **91**, 215001 (2003).

[112] P. McKenna, K.W.D. Ledingham, I. Spencer, T. McCany, R.P. Singhal, C. Ziener, P.S. Foster, E.J. Divall, C.J. Hooker, D. Neely, A.J. Langley, R.J. Clarke, P.A. Norreys, K. Krushelnick, and E.L. Clark. *Characterization of multiterawatt laser-solid interactions for proton acceleration*. Rev. Sci. Instrum. **73**, 4176 (2002).

[113] F. Lindau, O. Lundh, A. Persson, P. McKenna, K. Osvay, D. Batani, and C.-G. Wahlström. *Laser-accelerated protons with energy-dependent beam direction*. Phys. Rev. Lett. **95**, 175002 (2005).

[114] P. McKenna, F. Lindau, O. Lundh, D. Neely, A. Persson, and C.-G. Wahlström. *High-intensity laser-driven proton acceleration: influence of pulse contrast*. Philosophical transactions of the Royal Society Series A **364**, 711–23 (2006).

[115] O. Lundh, Y. Glinec, C. Homann, F. Lindau, A. Persson, C.-G. Wahlström, D.C. Carroll, and P. McKenna. *Active steering of laser-accelerated ion beams*. Appl. Phys. Lett. **92**, 1504 (2008).

[116] D. Neely, P. Foster, A. Robinson, F. Lindau, O. Lundh, A. Persson, C.-G. Wahlström, and P. McKenna. *Enhanced proton beams from ultrathin targets driven by high contrast laser pulses*. Appl. Phys. Lett. **89**, 1502 (2006).

[117] P. Antici, J. Fuchs, E. d'Humières, E. Lefebvre, M. Borghesi, E. Brambrink, C.A. Cecchetti, S. Gaillard, L. Romagnani, Y. Sentoku, T. Toncian, O. Willi, P. Audebert, and H. Pépin. *Energetic protons generated by ultrahigh contrast laser pulses interacting with ultrathin targets*. Phys. Plasmas **14**, 030701 (2007).

[118] T. Ceccotti, A. Lévy, H. Popescu, F. Réau, P. D'Oliveira, P. Monot, J.P. Geindre, E. Lefebvre, and Ph. Martin. *Proton acceleration with high-intensity ultrahigh-contrast laser pulses*. Phys. Rev. Lett **99**, 185002 (2007).

[119] M. Roth, E. Brambrink, P. Audebert, A. Blazevic, R. Clarke, J. Cobble, T.E. Cowan, J. Fernandez, J. Fuchs, M. Geissel, D. Habs, M. Hegelich, S. Karsch, K. Ledingham, D. Neely, H. Ruhl, T. Schlegel, and J. Schreiber. *Laser accelerated ions and electron transport in ultra-intense laser matter interaction*. Laser Part. Beams **23**, 95 (2005).

[120] A.P.L. Robinson, D. Neely, P. McKenna, and R.G. Evans. *Spectral control in proton acceleration with multiple laser pulses*. Plasma Phys. Control. Fusion **49**, 373 (2007).

[121] A.P.L. Robinson, M. Sherlock, and P.A. Norreys. *Artificial collimation of fast-electron beams with two laser pulses*. Phys. Rev. Lett. **100**, 025002 (2008).

[122] X.Q. Yan, C. Lin, Z.M. Sheng, Z.Y. Guo, B.C. Liu, Y.R. Lu, J.X. Fang, and J.E. Chen. *Generating high-current monoenergetic proton beams by a circularly polarized laser pulse in the phase-stable acceleration regime*. Phys. Rev. Lett. **100**, 135003 (2008).

[123] A.P.L. Robinson, M. Zepf, S. Kar, R.G. Evans, and C. Bellei. *Radiation pressure acceleration of thin foils with circularly polarized laser pulses*. New J. Phys. **10**, 013021 (2008).

[124] O. Klimo, J. Psikal, J. Limpouch, and V.T. Tikhonchuk. *Monoenergetic ion beams from ultrathin foils irradiated by ultrahigh-contrast circularly polarized laser pulses*. Physical Review Special Topics - Accelerators and Beams **11**, 031301 (2008).

[125] X. Zhang, B. Shen, X. Li, Z. Jin, F. Wang, and M. Wen. *Efficient GeV ion generation by ultraintense circularly polarized laser pulse.* Phys. Plasmas **14**, 123108 (2007).

[126] R.P. Johnson, N.K. Moncur, J.A. Cobble, R.G. Watt, and R.B. Gibson. *Trident as an ultrahigh irradiance laser.* Proc. SPIE **2377**, 294 (1995).

[127] J. Schreiber. *Ion acceleration driven by high-intensity laser pulses.* Dissertation, Ludwig-Maximilians-Universität München, Fakultät für Physik, Schellingstraße 4, 80799 München, Germany, (2006).

[128] J. Zou, D. Descamps, P. Audebert, S.D. Baton, J.L. Paillard, D. Pesme, A. Michard, A.M. Sautivet, H. Timsit, and A. Migus. *LULI 100-TW Ti:sapphire/Nd:glass laser: a first step toward a high-performance petawatt facility.* Proc. SPIE **3492**, 94 (1999).

[129] M. Roth, B. Becker-de Mos, R. Bock, S. Borneis, H. Brandt, C. Bruske, and J.A. Caird. *PHELIX: a petawatt high-energy laser for heavy ion experiments.* Proc. SPIE **4424**, 78 (2001).

[130] J. Schwarz, P. Rambo, M. Geissel, A. Edens, I. Smith, E. Brambrink, M. Kimmel, and B. Atherton. *Activation of the Z-petawatt laser at Sandia National Laboratories.* J. Phys.: Conf. Ser. **112**, 2020 (2008).

[131] P.J. Mohr and B.N. Taylor. *CODATA recommended values of the fundamental physical constants: 2002.* Rev. Mod. Phys. **77**, 1 – 107 (2005).

[132] W.L. Kruer. *The physics of laser plasma interactions.* Addison-Wesley, New York, (1988).

[133] P. Gibbon. *Short pulse laser interactions with matter.* Imperial College Press, London, (2005).

[134] S.C. Wilks and W.L. Kruer. *Absorption of ultrashort, ultra-intense laser light by solids and overdense plasma.* IEEE Journal of Quantum Electronics **33**, 1954 – 1968 (1997).

[135] D. Umstadter. *Relativistic laser-plasma interactions.* J. Phys. D Appl. Phys. **36**, R151–R165 (2003).

[136] E. Esarey, P. Sprangle, and J. Krall. *Laser acceleration of electrons in vacuum.* Phys. Rev. E **52**, 5443 – 5453 (1995).

[137] P. Drude. *Über Fernewirkungen.* Annalen der Physik **298**, 1 (1897).

[138] H. Boot and R.B.R Shersby-Harvie. *Charged Particles in a Non-uniform Radio-frequency Field.* Nature **180**, 1187 (1957).

[139] D. Bauer, P. Mulser, and W. Steeb. *Relativistic ponderomotive force, Uphill acceleration, and transition to chaos.* Phys. Rev. Lett. **75**, 4622–4625 (1995).

[140] P. Mora and T.M. Antonsen. *Kinetic modeling of intense, short laser pulses propagating in tenuous plasmas.* Phys. Plasmas **4**, 217 (1997).

[141] D.R. Bituk and M.V. Fedorov. *Relativistic ponderomotive forces.* Journal of Experimental and Theoretical Physics **89**, 640 (1999).

[142] F.V. Hartemann, S.N. Fochs, G.P. LeSage, N.C. Luhmann Jr., J.G. Woodworth, M.D. Perry, Y.J. Chen, and A.K. Kerman. *Nonlinear ponderomotive scattering of relativistic electrons by an intense laser field at focus.* Phys. Rev. E **51**, 4833–4843 (1995).

[143] B. Quesnel and P. Mora. *Theory and simulation of the interaction of ultraintense laser pulses with electron in vacuum.* Phys. Rev. E **58**, 3719–3732 (1998).

[144] D.D. Meyerhofer. *High-intensity-laser-electron scattering.* IEEE Journal of Quantum Electronics **33**, 1935 – 1941 (1997).

[145] P. Mulser, R. Sigel, and S. Witkowski. *Plasma production by laser.* Physics Reports **6**, 187 – 239 (1973).

[146] T. Tajima and J.M. Dawson. *Laser electron accelerator.* Phys. Rev. Lett. **43**, 267 (1979).

[147] A Pukhov and J Meyer-Ter-Vehn. *Laser wake field acceleration: the highly non-linear broken-wave regime.* Applied Physics B Lasers and Optics **74**, 355 (2002).

[148] S.P.D Mangles, C.D. Murphy, Z. Najmudin, A.G.R. Thomas, J.L. Collier, A.E. Dangor, E.J. Divall, P.S. Foster, J.G. Gallacher, C.J. Hooker, D.A. Jaroszynski, A.J. Langley, W.B. Mori, P.A. Norreys, F.S. Tsung, R. Viskup, B.R. Walton, and K. Krushelnick. *Monoenergetic beams of relativistic electrons from intense laser-plasma interactions.* Nature **431**, 535–8 (2004).

[149] C.G.R Geddes, C.S. Toth, J. Van Tilborg, E. Esarey, C.B. Schroeder, D. Bruhwiler, C. Nieter, J. Cary, and W.P. Leemans. *High-quality electron beams from a laser wakefield accelerator using plasma-channel guiding.* Nature **431**, 538–41 (2004).

[150] J. Faure, Y. Glinec, A. Pukhov, S. Kiselev, S. Gordienko, E. Lefebvre, J.-P. Rousseau, F. Burgy, and V. Malka. *A laserĐplasma accelerator producing monoenergetic electron beams.* Nature **431**, 541–544 (2004).

[151] J. Faure, C. Rechatin, A. Norlin, A. Lifschitz, Y. Glinec, and V. Malka. *Controlled injection and acceleration of electrons in plasma wakefields by colliding laser pulses.* Nature **444**, 737–9 (2006).

[152] W. P. Leemans, B. Nagler, A. J. Gonsalves, Cs. Toth, K. Nakamura, C. G. R. Geddes, E. Esarey, C. B. Schroeder, and S. M. Hooker. *GeV electron beams from a centimetre-scale accelerator.* Nature Phys. **2**, 696–9 (2006).

[153] K. Nakamura, B. Nagler, Cs. Tóth, C. G. R. Geddes, C. B. Schroeder, E. Esarey, W. P. Leemans, A. J. Gonsalves, and S. M. Hooker. *GeV electron beams from a centimeter-scale channel guided laser wakefield accelerator.* Phys. Plasmas **14**, 056708 (2007).

[154] S. Karsch, J. Osterhoff, A. Popp, T.P. Rowlands-Rees, Zs. Major, M. Fuchs, B. Marx, R. Hörlein, K. Schmid, L. Veisz, S. Becker, U. Schramm, B. Hidding, G. Pretzler, D. Habs, F. Grüner, F. Krausz, and S.M. Hooker. *GeV-scale electron acceleration in a gas-filled capillary discharge waveguide.* New J. Phys. **9**, 415 (2007).

[155] F. Brunel. *Not-so-resonant, resonant absorption.* Phys. Rev. Lett. **59**, 52–5 (1987).

[156] W.L. Kruer and K. Estabrook. $j \times B$ *heating by very intense laser light.* Phys. Fluids **28**, 430 (1985).

[157] S. Wilks, W. Kruer, M. Tabak, and A. Langdon. *Absorption of ultra-intense laser pulses.* Phys. Rev. Lett. **69**, 1383–1386 (1992).

[158] G. Malka and J.L. Miquel. *Experimental Confirmation of Ponderomotive-Force Electrons Produced by an Ultrarelativistic Laser Pulse on a Solid Target.* Phys. Rev. Lett. **77**, 75 (1996).

[159] S.D. Baton, J.J. Santos, F. Amiranoff, H. Popescu, L. Gremillet, M. Koenig, E. Martinolli, O. Guilbaud, C. Rousseaux, M. RabecLeGloahec, T. Hall, D. Batani, E. Perelli, F. Scianitti, and T.E. Cowan. *Evidence of Ultrashort Electron Bunches in Laser-Plasma Interactions at Relativistic Intensities.* Phys. Rev. Lett. **91**, 105001 (2003).

[160] J. Zheng, K.A. Tanaka, T. Sato, T. Yabuuchi, T. Kurahashi, Y. Kitagawa, R. Kodama, T. Norimatsu, and T. Yamanaka. *Study of hot electrons by measurement of optical emission from the rear surface of a metallic foil irradiated with ultraintense laser pulse.* Phys. Rev. Lett. **92**, 165001 (2004).

[161] H Popescu, S. D Baton, F Amiranoff, C Rousseaux, M. Rabec Le Gloahec, J. J Santos, L Gremillet, M Koenig, E Martinolli, T Hall, J. C Adam, A Heron, and D Batani. *Subfemtosecond, coherent, relativistic, and ballistic electron bunches generated at ω_0 and $2\omega_0$ in high intensity laser-matter interaction.* Phys. Plasmas **12**, 063106 (2005).

[162] J.J. Santos, A. Debayle, Ph. Nicolaï, V. Tikhonchuk, M. Manclossi, D. Batani, A. Guemnie-Tafo, J. Faure, V. Malka, and J.J. Honrubia. *Fast-electron transport and induced heating in aluminum foils.* Phys. Plasmas **14**, 103107 (2007).

[163] D. Bauer and P. Mulser. *Vacuum heating versus skin layer absorption of intense femtosecond laser pulses.* Phys. Plasmas **14**, 023301 (2007).

[164] B. Bezzerides, S. Gitomer, and D. Forslund. *Randomness, Maxwellian Distributions, and Resonance Absorption.* Phys. Rev. Lett. **44**, 651 – 654 (1980).

[165] F. Jüttner. *Das Maxwellsche Gesetz der Geschwindigkeitsverteilung in der Relativtheorie.* Annalen der Physik **339**, 856 – 82 (1911).

[166] P.A. Norreys, M. Santala, E. Clark, M. Zepf, I. Watts, F.N. Beg, K. Krushelnick, M. Tatarakis, A.E. Dangor, X. Fang, P. Graham, T. McCanny, R.P. Singhal, K.W.D. Ledingham, A. Creswell, D.C.W. Sanderson, J. Magill, A. Machacek, J.S. Wark, R. Allott, B. Kennedy, and D. Neely. *Observation of a highly directional γ-ray beam from ultrashort, ultraintense laser pulse interactions with solids.* Phys. Plasmas **6**, 2150 (1999).

[167] M.H. Key, M.D. Cable, T.E. Cowan, K.G. Estabrook, B.A. Hammel, S.P. Hatchett, E.A. Henry, D.E. Hinkel, J.D. Kilkenny, J.A. Koch, W.L. Kruer, A.B. Langdon, B.F. Lasinski, R.W. Lee, B.J. MacGowan, A. MacKinnon, J.D. Moody, M.J. Moran, A.A. Offenberger, D.M. Pennington, M.D. Perry, T.J. Phillips, T.C. Sangster, M.S. Singh, M.A. Stoyer, M. Tabak, G.L. Tietbohl, M. Tsukamoto, K. Wharton, and S.C. Wilks. *Hot electron production and heating by hot electrons in fast ignitor research.* Phys. Plasmas **5**, 1966 (1998).

[168] D. Batani, R.R. Freeman, and S. Baton. *Progress in ultrafast intense laser sciene III.* Springer-Verlag, Heidelberg, springer series in chemical physics 89 edition, (2008).

[169] J.R. Davies. *How wrong is collisional Monte Carlo modeling of fast electron transport in high-intensity laser-solid interactions?* Phys. Rev. E **65**, 026407 (2002).

[170] T. Feurer, W. Theobald, R. Sauerbrey, I. Uschmann, D. Altenbernd, U. Teubner, P. Gibbon, E. Förster, G. Malka, and J.L. Miquel. *Onset of diffuse reflectivity and fast electron flux inhibition in 528-nm-laser-solid interactions at ultrahigh intensity.* Phys. Rev. E **56**, 4608 – 14 (1997).

[171] Y Ping, R Shepherd, B. F Lasinski, M Tabak, H Chen, H. K Chung, K. B Fournier, S. B Hansen, A Kemp, D. A Liedahl, K Widmann, S. C Wilks, W Rozmus, and M Sherlock. *Absorption of short laser pulses on solid targets in the ultrarelativistic regime.* Phys. Rev. Lett. **100**, 085004 (2008).

[172] Hui Chen and Scott C. Wilks. *Evidence of enhanced effective hot electron temperatures in ultraintense laser-solid interactions due to reflexing.* Laser Part. Beams **23**, 411 (2005).

[173] F. Brandl, G. Pretzler, D. Habs, and E. Fill. *Čerenkov radiation diagnostics of hot electrons generated by fs-laser interaction with solid targets.* Europhysics Letters **61**, 632 – 638 (2003).

[174] T.E. Cowan, A.W. Hunt, T.W. Phillips, S.C. Wilks, M.D. Perry, C. Brown, W. Fountain, S. Hatchett, J. Johnson, M.H. Key, T. Parnell, D.M. Pennington, R.A. Snavely, and Y. Takahashi. *Photonuclear Fission from High Energy Electrons from Ultraintense Laser-Solid Interactions.* Phys. Rev. Lett. **84**, 903 (2000).

[175] J.J. Santos, F. Amiranoff, S.D. Baton, L. Gremillet, M. Koenig, E. Martinolli, M. Rabec Le Gloahec, C. Rousseaux, D. Batani, A. Bernardinello, G. Greison, and T. Hall. *Fast Electron Transport in Ultraintense Laser Pulse Interaction with Solid Targets by Rear-Side Self-Radiation Diagnostics.* Phys. Rev. Lett. **89**, 025001 (2002).

[176] F.N. Beg, A.R. Bell, A.E. Dangor, C.N. Danson, A.P. Fews, M.E. Glinsky, B.A. Hammel, P. Lee, P.A. Norreys, and M. Tatarakis. *A study of picosecond laser-solid interactions up to 10^{19} Wcm^{-2}.* Phys. Plasmas **4**, 447 (1997).

[177] Erik Lefebvre and Guy Bonnaud. *Nonlinear electron heating in ultrahigh-intensity-laser-plasma interaction.* Phys. Rev. E **55**, 1011 (1997).

[178] J. Fuchs, J.C. Adam, F. Amiranoff, S.D. Baton, P. Gallant, L. Gremillet, A. Héron, J.C. Kieffer, G. Laval, G. Malka, J.L. Miquel, P. Mora, H. Pépin, and C. Rousseaux. *Transmission through highly overdense plasma slabs with a subpicosecond relativistic laser pulse.* Phys. Rev. Lett. **80**, 2326 (1998).

[179] M.I. Santala, M. Zepf, I. Watts, F.N. Beg, E. Clark, M. Tatarakis, K. Krushelnick, A.E. Dangor, T. Mccanny, I. Spencer, R.P. Singhal, K.W. Ledingham, S.C. Wilks, A.C. Machacek, J.S. Wark, R. Allott, R.J. Clarke, and P.A. Norreys. *Effect of the Plasma Density Scale Length on the Direction of Fast Electrons in Relativistic Laser-Solid Interactions.* Phys. Rev. Lett. **84**, 1459 (2000).

[180] K. Adumi, K.A. Tanaka, T. Matsuoka, T. Kurahashi, T. Yabuuchi, Y. Kitagawa, R. Kodama, K. Sawai, K. Suzuki, K. Okabe, T. Sera, T. Norimatsu, and Y. Izawa. *Characterization of preplasma produced by an ultrahigh intensity laser system.* Phys. Plasmas **11**, 3721 (2004).

[181] R. Kodama, K. Takahashi, K.A. Tanaka, M. Tsukamoto, H. Hashimoto, Y. Kato, and K. Mima. *Study of Laser-Hole Boring into Overdense Plasmas*. Phys. Rev. Lett. **77**, 4906 (1996).

[182] J.S. Green, V.M. Ovchinnikov, R.G. Evans, K.U. Akli, H. Azechi, F.N. Beg, C. Bellei, R.R. Freeman, H. Habara, R. Heathcote, M.H. Key, J.A. King, K.L. Lancaster, N.C. Lopes, T. Ma, A.J. Mackinnon, K. Markey, A. Mcphee, Z. Najmudin, P. Nilson, R. Onofrei, R. Stephens, K. Takeda, K.A. Tanaka, W. Theobald, T. Tanimoto, J. Waugh, L. Van Woerkom, N.C. Woolsey, M. Zepf, J.R. Davies, and P.A. Norreys. *Effect of laser intensity on fast-electron-beam divergence in solid-density plasmas*. Phys. Rev. Lett. **100**, 015003 (2008).

[183] B. Hidding, G. Pretzler, M. Clever, F. Brandl, F. Zamponi, A. Lübcke, T. Kämpfer, I. Uschmann, E. Förster, U. Schramm, R. Sauerbrey, E. Kroupp, L. Veisz, K. Schmid, S. Benavides, and S. Karsch. *Novel method for characterizing relativistic electron beams in a harsh laser-plasma environment*. Rev. Sci. Instrum. **78**, 083301 (2007).

[184] M Nakatsutsumi, R Kodama, P. A Norreys, S Awano, H Nakamura, T Norimatsu, A Ooya, M Tampo, K. A Tanaka, T Tanimoto, T Tsutsumi, and T Yabuuchi. *Reentrant cone angle dependence of the energetic electron slope temperature in high-intensity laser-plasma interactions*. Phys. Plasmas **14**, 050701 (2007).

[185] H. Teng, J. Zhang, Z.L. Chen, Y.T. Li, X. Lu, K. Li, and X.Y. Peng. *Hot-electron-induced plasma formation on the rear surface of a foil*. Appl. Phys. B **76**, 687 (2003).

[186] G Schaumann, **M Schollmeier**, G Rodriguez-Prieto, A Blazevic, E Brambrink, M Geissel, S Korostiy, P Pirzadeh, M Roth, F B Rosmej, A Ya Faenov, T A Pikuz, K Tsigutkin, Y Maron, N A Tahir, and D H H Hoffmann. *High energy heavy ion jets emerging from laser plasma generated by long pulse laser beams from the NHELIX laser system at GSI*. Laser Part. Beams **23**, 503 – 512 (2005).

[187] T. Ditmire, T. Donnelly, A.M. Rubenchik, R.W. Falcone, and M.D. Perry. *Interaction of intense laser pulses with atomic clusters*. Phys. Rev. A **53**, 3379 (1996).

[188] K. Krushelnick, E.L. Clark, Z. Najmudin, M. Salvati, M.I.K. Santala, M. Tatarakis, A.E. Dangor, V. Malka, D. Neely, R. Allott, and C. Danson. *Multi-MeV Ion Production from High-Intensity Laser Interactions with Underdense Plasmas*. Phys. Rev. Lett. **83**, 737 (1999).

[189] A. V Kuznetsov, T. Zh Esirkepov, F. F Kamenets, and S. V Bulanov. *Efficiency of Ion Acceleration by a Relativistically Strong Laser Pulse in an Underdense Plasma*. Plasma Physics Reports **27**, 211 (2001).

[190] H. Habara, K.L. Lancaster, S. Karsch, C.D. Murphy, P.A. Norreys, R.G. Evans, M. Borghesi, L. Romagnani, M. Zepf, T. Norimatsu, Y. Toyama, R. Kodama, J.A. King, R. Snavely, K. Akli, B. Zhang, R. Freeman, S. Hatchett, A.J. Mackinnon, P. Patel, M.H. Key, C. Stoeckl, R.B. Stephens, R.A. Fonseca, and L.O. Silva. *Ion acceleration from the shock front induced by hole boring in ultraintense laser-plasma interactions*. Phys. Rev. E **70**, 46414 (2004).

[191] J. Badziak, S. Glowacz, S. Jablonski, P. Parys, J. Wolowski, H. Hora, J. Krása, L. Láska, and K. Rohlena. *Production of ultrahigh ion current densities at skin-layer subrelativistic laser plasma interaction*. Plasma Phys. Control. Fusion **46**, 541 (2004).

[192] J. Badziak, P. Antici, J. Fuchs, S. Jablowski, L. Lancia, A. Mancic, P. Parys, M. Rosiński, R. Suchańska, A. Szydlowski, and J. Wolowski. *Ultrahigh-current proton beams from short-pulse laser-solid interactions.* J. Phys.: Conf. Ser. **112**, 042040 (2008).

[193] L. Yin, B.J. Albright, B.M. Hegelich, and J.C. Fernández. *GeV laser ion acceleration from ultrathin targets: The laser break-out afterburner.* Laser Part. Beams **24**, 291 (2006).

[194] T. Esirkepov, M. Borghesi, S.V. Bulanov, G. Mourou, and T. Tajima. *Highly Efficient Relativistic-Ion Generation in the Laser-Piston Regime.* Phys. Rev. Lett. **92**, 175003 (2004).

[195] C.I. Moore, J.P. Knauer, and D.D. Meyerhofer. *Observation of the Transition from Thomson to Compton Scattering in Multiphoton Interactions with Low-Energy Electrons.* Phys. Rev. Lett. **74**, 2439 (1995).

[196] A.R. Bell, J.R. Davies, S. Guerin, and H. Ruhl. *Fast-electron transport in high-intensity short-pulse laser-solid experiments.* Plasma Phys. Control. Fusion **39**, 653 (1997).

[197] A.R. Bell, A.P.L. Robinson, M. Sherlock, R.J. Kingham, and W. Rozmus. *Fast electron transport in laser-produced plasmas and the KALOS code for solution of the Vlasov-Fokker-Planck equation.* Plasma Phys. Control. Fusion **48**, 37 (2006).

[198] E. Martinolli, M. Koenig, S.D. Baton, J.J. Santos, F. Amiranoff, D. Batani, E. Perelli-Cippo, F. Scianitti, L. Gremillet, R. Mélizzi, A. Decoster, C. Rousseaux, T.A. Hall, M.H. Key, R. Snavely, A.J. Mackinnon, R.R. Freeman, J.A. King, R. Stephens, D. Neely, and R.J. Clarke. *Fast-electron transport and heating of solid targets in high-intensity laser interactions measured by K_α fluorescence.* Phys. Rev. E **73**, 046402 (2006).

[199] F. Pisani, A. Bernardinello, D. Batani, A. Antonicci, E. Martinolli, M. Koenig, L. Gremillet, F. Amiranoff, S. Baton, J. Davies, T. Hall, D. Scott, P. Norreys, A. Djaoui, C. Rousseaux, P. Fews, H. Bandulet, and H. Pepin. *Experimental evidence of electric inhibition in fast electron penetration and of electric-field-limited fast electron transport in dense matter.* Phys. Rev. E **62**, R5927 (2000).

[200] H. Nishimura, Y. Inubushi, M. Ochiai, T. Kai, T. Kawamura, S. Fujioka, M. Hashida, S. Simizu, S. Sakabe, R. Kodama, K.A. Tanaka, S. Kato, F. Koike, S. Nakazaki, H. Nagatomo, T. Johzaki, and K. Mima. *Study of fast electron transport in hot dense matter using x-ray spectroscopy.* Plasma Phys. Control. Fusion **47**, 823 (2005).

[201] F. Califano, F. Pegoraro, and S.V. Bulanov. *Spatial structure and time evolution of the Weibel instability in collisionless inhomogeneous plasmas.* Phys. Rev. E **56**, 963 (1997).

[202] D. Batani, S.D. Baton, M. Manclossi, J.J. Santos, F. Amiranoff, M. Koenig, E. Martinolli, A. Antonicci, C. Rousseaux, M. Rabec Le Gloahec, .T Hall, V. Malka, T.E. Cowan, J. King, R.R. Freeman, M. Key, and R. Stephens. *Ultraintense Laser-Produced Fast-Electron Propagation in Gas Jets.* Phys. Rev. Lett. **94**, 055004 (2005).

[203] M.S. Wei, F.N. Beg, E.L. Clark, A.E. Dangor, R.G. Evans, A. Gopal, K.W. Ledingham, P. McKenna, P.A. Norreys, M. Tatarakis, M. Zepf, and K. Krushelnick. *Observations of the filamentation of high-intensity laser-produced electron beams.* Phys. Rev. E **70**, 056412 (2004).

[204] H. Ruhl. *Collective super-intense laser-plasma interaction.* Habilitationsschrift, Technische Universität Darmstadt, Theoretical Quantum Electronics, Institut für Angewandte Physik, Schlossgartenstraße 7, 64289 Darmstadt, Germany, (2000).

[205] H. Ruhl, A. Macchi, P. Mulser, F. Cornolti, and S. Hain. *Collective dynamics and enhancement of absorption in deformed targets.* Phys. Rev. Lett. **82**, 2095 (1999).

[206] M. Honda and J. Meyer-Ter-Vehn and. A Pukhov. *Collective stopping and ion heating in relativistic-electron-beam transport for Fast Ignition.* Phys. Rev. Lett. **85**, 2128 (2000).

[207] Y. Sentoku, K. Mima, Z.M. Sheng, P. Kaw, K. Nishihara, and K. Nishikawa. *Three-dimensional particle-in-cell simulations of energetic electron generation and transport with relativistic laser pulses in overdense plasmas.* Phys. Rev. E **65**, 046408 (2002).

[208] L. Gremillet, G. Bonnaud, and F. Amiranoff. *Filamented transport of laser-generated relativistic electrons penetrating a solid target.* Phys. Plasmas **9**, 941 (2002).

[209] M. Borghesi, D.H. Campbell, A. Schiavi, M.G. Haines, O. Willi, A.J. MacKinnon, P. Patel, L.A. Gizzi, M. Galimberti, R.J. Clarke, F. Pegoraro, H. Ruhl, and S. Bulanov. *Electric field detection in laser-plasma interaction experiments via the proton imaging technique.* Phys. Plasmas **9**, 2214 (2002).

[210] H. Alfvén. *On the Motion of Cosmic Rays in Interstellar Space.* Physical Review **55**, 425 (1939).

[211] J. R Davies. *The Alfvén limit revisited and its relevance to laser-plasma interactions.* Laser Part. Beams **24**, 299 (2006).

[212] A Pukhov and J Meyer-Ter-Vehn. *Laser Hole Boring into Overdense Plasma and Relativistic Electron Currents for Fast Ignition of ICF Targets.* Phys. Rev. Lett. **79**, 2686 (1997).

[213] R.B. Stephens, R.A. Snavely, Y. Aglitskiy, F. Amiranoff, C. Andersen, D. Batani, S.D. Baton, T. Cowan, R.R. Freeman, T. Hall, S.P. Hatchett, J.M. Hill, M.H. Key, J.A. King, J.A. Koch, M. Koenig, A.J. Mackinnon, K.L. Lancaster, E. Martinolli, P. Norreys, E. Perelli-Cippo, M. Rabec Le Gloahec, C. Rousseaux, J.J. Santos, and F. Scianitti. *K_α fluorescence measurement of relativistic electron transport in the context of fast ignition.* Phys. Rev. E **69**, 066414 (2004).

[214] M. Manclossi, J.J. Santos, D. Batani, J. Faure, A. Debayle, V.T. Tikhonchuk, and V. Malka. *Study of Ultraintense Laser-Produced Fast-Electron Propagation and Filamentation in Insulator and Metal Foil Targets by Optical Emission Diagnostics.* Phys. Rev. Lett. **96**, 125002 (2006).

[215] L. Gremillet, F. Amiranoff, S.D. Baton, J.-C. Gauthier, M. Koenig, E. Martinolli, F. Pisani, G. Bonnaud, C. Lebourg, C. Rousseaux, C. Toupin, A. Antonicci, D. Batani, A. Bernardinello, T. Hall, D. Scott, P. Norreys, H. Bandulet, and H. Pépin. *Time-resolved observation of ultrahigh intensity laser-produced electron jets propagating through transparent solid targets.* Phys. Rev. Lett. **83**, 5015 (1999).

[216] P.A. Norreys, J.S. Green, J.R. Davies, M. Tatarakis, E.L. Clark, F.N. Beg, A.E. Dangor, K.L. Lancaster, M.S. Wei, M. Zepf, and K. Krushelnick. *Observation of annular electron*

beam transport in multi-TeraWatt laser-solid interactions. Plasma Phys. Control. Fusion **48**, L11 (2006).

[217] J.A. Koch, M.H. Key, R.R. Freeman, S.P. Hatchett, R.W. Lee, D. Pennington, R.B. Stephens, and M. Tabak. *Experimental measurements of deep directional columnar heating by laser-generated relativistic electrons at near-solid density.* Phys. Rev. E **65**, 016410 (2002).

[218] S.I. Krasheninnikov, A.V. Kim, B.K. Frolov, and R. Stephens. *Intense electron beam propagation through insulators: Ionization front structure and stability.* Phys. Plasmas **12**, 073105 (2005).

[219] A.R. Bell and R.J. Kingham. *Resistive Collimation of Electron Beams in Laser-Produced Plasmas.* Phys. Rev. Lett. **91**, 035003 (2003).

[220] J.R. Davies, A.R. Bell, and M. Tatarakis. *Magnetic focusing and trapping of high-intensity laser-generated fast electrons at the rear of solid targets.* Phys. Rev. E **59**, 6032 (1999).

[221] J. Honrubia, A. Antonicci, and D. Moreno. *Hybrid simulations of fast electron transport in conducting media.* Laser Part. Beams **22**, 129 – 35 (2004).

[222] J.J. Honrubia, M. Kaluza, J. Schreiber, G.D. Tsakiris, and J. Meyer-Ter-Vehn. *Laser-driven fast-electron transport in preheated foil targets.* Phys. Plasmas **12**, 052708 (2005).

[223] J.J. Honrubia, C. Alfonsin, L. Alonso, B. Perez, and J.A. Cerrada. *Simulations of heating of solid targets by fast electrons.* Laser Part. Beams **24**, 217 – 22 (2006).

[224] K.L. Lancaster, J.S. Green, D.S. Hey, K.U. Akli, J.R. Davies, R.J. Clarke, R.R. Freeman, H. Habara, M.H. Key, R. Kodama, K. Krushelnick, C.D. Murphy, M. Nakatsutsumi, P. Simpson, R. Stephens, C. Stoeckl, T. Yabuuchi, M. Zepf, and P.A. Norreys. *Measurements of Energy Transport Patterns in Solid Density Laser Plasma Interactions at Intensities of 5×10^{20} Wcm^{-2}.* Phys. Rev. Lett. **98**, 125002 (2007).

[225] H.A. Bethe. *Molière's Theory of Multiple Scattering.* Physical Review **89**, 1256 (1953).

[226] W.T. Scott. *The Theory of Small-Angle Multiple Scattering of Fast Charged Particles.* Reviews of Modern Physics **35**, 231 (1963).

[227] Geant4 Collaboration. *GEANT4 - a simulation toolkit.* Nucl. Instr. and Meth. A **506**, 250 (2003).

[228] M. Günther. Dissertation, TU Darmstadt, Institut für Kernphysik, Schlossgartenstraße. 9, 64289 Darmstadt, Germany, *(in preparation)*.

[229] K.U. Akli, M.H. Key, H.K. Chung, S.B. Hansen, R.R. Freeman, M.H. Chen, G. Gregori, S. Hatchett, D. Hey, N. Izumi, J. King, J. Kuba, P. Norreys, A.J. Mackinnon, C.D. Murphy, R. Snavely, R.B. Stephens, C. Stoeckel, W. Theobald, and B. Zhang. *Temperature sensitivity of Cu K_α imaging efficiency using a spherical Bragg reflecting crystal.* Phys. Plasmas **14**, 023102 (2007).

[230] J.E. Crow, P.L. Auer, and J.E. Allen. *The expansion of a plasma into a vacuum.* J. Plasma Physics **14**, 65 – 76 (1975).

[231] N.A. Krall and A.W. Trivelpiece. *Principles of Plasma Physics.* San Francisco Press, Inc., San Francisco, (1986).

[232] S. Augst, D. Strickland, D.D. Meyerhofer, S.L. Chin, and J.H. Eberly. *Tunneling ionization of noble gases in a high-intensity laser field.* Phys. Rev. Lett. **63**, 2212 (1989).

[233] The Mathworks Inc. *MATLAB.* Website, June (2008). Available online at http://www.mathworks.com/ (2008).

[234] M. Passoni and M. Lontano. *One-dimensional model of the electrostatic ion acceleration in the ultraintense laser solid interaction.* Laser Part. Beams **22**, 163 (2004).

[235] S.V. Bulanov, T.Zh. Esirkepov, J. Koga, T. Tajima, and D. Farina. *Concerning the maximum energy of ions accelerated at the front of a relativistic electron cloud expanding into vacuum.* Plasma Physics Reports **30**, 18 (2004).

[236] A. Maksimchuk, K. Flippo, H. Krause, G. Mourou, K. Nemoto, D. Shultz, D. Umstadter, R. Vane, V.Yu. Bychenkov, G.I. Dudnikova, V.F. Kovalev, K. Mima, V.N. Novikov, Y. Sentoku, and S.V. Tolokonnikov. *High-energy ion generation by short laser pulses.* Plasma Physics Reports **30**, 473 (2004).

[237] S. Betti, F. Ceccherini, F. Cornolti, and F. Pegoraro. *Expansion of a finite-size plasma in vacuum.* Plasma Phys. Control. Fusion **47**, 521 (2005).

[238] N. Kumar and A. Pukhov. *Self-similar quasineutral expansion of a collisionless plasma with tailored electron temperature profile.* Phys. Plasmas **15**, 053103 (2008).

[239] M. Murakami and M.M. Basko. *Self-similar expansion of finite-size non-quasi-neutral plasmas into vacuum: Relation to the problem of ion acceleration.* Phys. Plasmas **13**, 012105 (2006).

[240] L.D. Landau and E.M. Lifshitz. *Mekhanika sploshnykh sred (Mechanics of Continous Media).* Gostekhizdat, (1954).

[241] S. Eliezer. *The interaction of high-power lasers with plasmas.* Institute of Physics Publishing, Bristol and Philadelphia, (2002).

[242] Malte Christoph Kaluza. *Characterisation of Laser-Accelerated Proton Beams.* PhD thesis, Technische Universität München, Fakultät für Physik, Schellingstraße 4, 80799 München, Germany, (2004).

[243] Erik Brambrink. *Untersuchung der Eigenschaften lasererzeugter Ionenstrahlen.* PhD thesis, Technische Universität Darmstadt, Institut für Kernphysik, Schlossgartenstraße 9, 64289 Darmstadt, Germany, (2004).

[244] A.V. Gurevich, L.V. Pariĭskaya, and L.P. Pitaevskiĭ. *Self-similar motion of rarefied plasma.* Soviet Physics JETP **22**, 449 (1966).

[245] M. Passoni, V.T. Tikhonchuk, M. Lontano, and V.Yu. Bychenkov. *Charge separation effects in solid targets and ion acceleration with a two-temperature electron distribution.* Phys. Rev. E **69**, 026411 (2004).

[246] V.Yu. Bychenkov, V.N. Novikov, D. Batani, V.T. Tikhonchuk, and S.G. Bochkarev. *Ion acceleration in expanding multispecies plasmas.* Phys. Plasmas **11**, 3242 (2004).

[247] P. Neumayer, R. Bock, S. Borneis, E. Brambrink, H. Brand, J. Caird, E.M. Campbell, E. Gaul, S. Goette, C. Haefner, T. Hahn, H.M. Heuck, D.H.H. Hoffmann, D. Javorkova, H.-J. Kluge, T. Kuehl, S. Kunzer, T. Merz, E. Onkels, M.D. Perry, D. Reemts, M. Roth, S. Samek, G. Schaumann, F. Schrader, W. Seelig, A. Tauschwitz, R. Thiel, D. Ursescu, P. Wiewior, U. Wittrock, and B. Zielbauer. *Status of PHELIX laser and first experiments.* Laser Part. Beams **23**, 385–389 (2005).

[248] K. Harres. *Experimentelle Bestimmung des Energiespektrums von lasererzeugten MeV-Ionenstrahlen mit Hilfe einer Thomson-Parabel.* Diplomarbeit, Technische Universität Darmstadt, Institut für Kernphysik, Schlossgartenstr. 9, 64289 Darmstadt, Germany, (2006).

[249] D. Nichiporov, V. Kostjuchenko, J.M. Puhl, D.L. Bensen, M.F. Desrosiers, C.E. Dick, W.L. McLaughlin, T. Kojima, B.M. Coursey, and S. Zink. *Investigation of applicability of alanine and radiochromic detectors to dosimetry of proton clinical beams.* Appl. Radiat. Isot. **46**, 1335 – 1362 (1995).

[250] S.M. Vatnitsky. *Radiochromic film dosimetry for clinical proton beams.* Appl. Radiat. Isot. **48**, 643 – 651 (1997).

[251] ISP - International Specialty Products, Wayne, New Jersey, USA. *Radiochromic film types HD-810, HS, MD-55, MD-V2-55 are trademarks of ISP corporation.* Website, June (2008). Available online at http://online1.ispcorp.com/`layouts/Gafchromic/index.html (2008).

[252] F. Nürnberg. *Vollständige Rekonstruktion und Transportsimulation eines laserbeschleunigten Protonenstrahls unter Verwendung von mikrostrukturierten Targetfolien und radiochromatischen Filmdetektoren.* Diplomarbeit, Technische Universität Darmstadt, Institut für Kernphysik, Schlossgartenstr. 9, 64289 Darmstadt, Germany, (2006).

[253] D. Kraus. *Kalibrierung radiochromatischer Filme zur Anwendung in der Laser-Ionenbeschleunigung.* Bachelor thesis, Technische Universität Darmstadt, Institut für Kernphysik, Schlossgartenstr. 9, 64289 Darmstadt, Germany, (2007).

[254] N.V. Klassen, L. van der Zwan, and J. Cygler. *GafChromic MD-55: Investigated as a precision dosimeter.* Med. Phys. **24**, 1924 – 1934 (1997).

[255] W.L. McLaughlin, M. Al-Sheikhly, D.F. Lewis, A. Kovács, and L. Wojnárovits. *Radiochromic solid-state polymerization reaction,* chapter 11, 152 – 166. ACS Symposium Series, edited by R.L. Clough and S.W. Shalaby. American Chemical Society, 620 edition (1996).

[256] M.J. Butson, P.K.N. Yu, T. Cheung, and P. Metcalfe. *Radiochromic film for medical radiation dosimetry.* Materials Science and Engineering: R: Reports **41**, 61–120 (2003).

[257] W.P. Bishop, K.C. Humpherys, and P.T. Randtke. *Poly(halo)styrene thin-film dosimeters for high doses.* Rev. Sci. Instrum. **44**, 443 – 452 (1973).

[258] A. Niroomand-Rad, C.R. Blackwell, B.M. Coursey, K.P. Gall, J.M. Galvin, W.L. McLaughlin, A.S. Meigooni, R. Nath, J.E. Rodgers, and C.G. Soares. *Radiochromic film dosimetry: Recommendations of AAPM radiation therapy committee task group 55.* Med. Phys. **25**, 2093 – 2115 (1998).

[259] W.L. McLaughlin, C. Yun-Dong, C.G. Soares, A. Miller, G. van Dyk, and D.F. Lewis. *Sensitometry of the response of a new radiochromic film dosimeter to gamma radiation and electron beams.* Nucl. Instr. and Meth. A **302**, 165 – 176 (1991).

[260] W.L. McLaughlin, M. Al-Sheikhly, D.F. Lewis, A. Kovács, and L. Wojnárovits. *A radiochromic solid-state polymerization reaction.* Polymer Preprints **35**, 920 – 921 (1994).

[261] H. Alva, H. Mercado-Uribe, M. Rodriguez-Villafuerte, and M.E. Brandan. *The use of a reflective scanner to study radiochromic film response.* Phys. Med. Biol. **47**, 2925 – 2933 (2002).

[262] G.R. Gluckman and L.E. Reinstein. *Comparison of three high-resolution digitizers for radiochromic film dosimetry.* Med. Phys. **29**, 1839 – 1846 (2002).

[263] D.S. Hey, M.H. Key, A.J. Mackinnon, A.G. MacPhee, P.K. Patel, R.R. Freeman, L.D. Van Woerkom, and C.M. Castaneda. *Use of GafChromic film to diagnose laser generated proton beams.* Rev. Sci. Instrum. **79**, 053501 (2008).

[264] A.M. Srivastava and C.R. Ronda. *Phosphors.* The Electrochemical Society Interface **Summer 2003**, 48–51 (2003).

[265] Stouffer Industries, Inc., 922 S. Cleveland Street, Mishawaka, IN 46544, USA. *Transparent step wedge #250915, part no. T4110cc.* Website, June (2008). Available online at http://www.stouffer.net (2008).

[266] Dmitry Varentsov. *Energy loss dynamics of intense heavy ion beams interacting with matter.* PhD thesis, Technische Universität Darmstadt, Institut für Kernphysik, Schlossgartenstraße 9, 64289 Darmstadt, Germany, (2002).

[267] J.P. Biersack J.F. Ziegler and U. Littmark. *The Stopping and Range of Ions in Matter.* Pergamon Press, New York, (1985). Available online at http://www.srim.org. (2008).

[268] B. Spielberger, M. Scholz, M. Krämer, and G. Kraft. *Experimental investigations of the response of films to heavy-ion irradiation.* Physics in Medicine and Biology **46**, 2889–97 (2001).

[269] B. Spielberger, M. Scholz, M. Krämer, and G. Kraft. *Calculation of the x-ray film response to heavy charged particle irradiation.* Physics in Medicine and Biology **47**, 4107–20 (2002).

[270] B. Spielberger, M. Krämer, and G. Kraft. *Three-dimensional dose verification with x-ray films in conformal carbon ion therapy.* Physics in Medicine and Biology **48**, 497–505 (2003).

[271] A. Piermattei, R. Miceli, L. Azario, A. Fidanzio, S. delle Canne, C. De Angelis, S. Onori, M. Pacilio, E. Petetti, L. Raffaele, and M.G. Sabini. *Radiochromic film dosimetry of a low energy proton beam.* Med. Phys. **27**, 1655–60 (2000).

[272] Institut für Mikroverfahrenstechnik, Forschungszentrum Karlsruhe, PF 3640, 76021 Karlsruhe, Germany. *Gruppe Mikrofluidische Komponenten & Fertigungsentwicklung*, June (2005). Mail address: Hermann-von-Helmholtz-Platz 1, 76344 Eggenstein-Leopoldshafen.

[273] Richard Romminger, Edelstahltechnik und Feinlaserschneiden. Available online at http://www.romminger.com (2008). Mail address: Gottlob-Bauknecht Straße 26, 75365 Calw, Germany.

[274] LFM - Laboratory for Precision Machining. Available online at http://www.lfm.uni-bremen (2008). Mail address: Badgasteiner Straße 2, 28359 Bremen, Germany.

[275] S.-W. Bahk, P. Rousseau, T.A. Planchon, V. Chvykov, G. Kalintchenko, A. Maksimchuk, G.A. Mourou, and V. Yanovsky. *Characterization of focal field formed by a large numerical aperture paraboloidal mirror and generation of ultra-high intensity (10^{22} W/cm^2)*. Appl. Phys. B **80**, 823 (2005).

[276] M. Zepf, E.L. Clark, K. Krushelnick, F.N. Beg, C. Escoda, A.E. Dangor, M.I.K. Santala, M. Tatarakis, I.F. Watts, P.A. Norreys, R.J. Clarke, J.R. Davies, M.A. Sinclair, R.D. Edwards, T.J. Goldsack, I. Spencer, and K.W.D. Ledingham. *Fast particle generation and energy transport in laser-solid interactions*. Phys. Plasmas **8**, 2323 (2001).

[277] M. Nishiuchi, A. Fukumi, H. Daido, Z. Li, A. Sagisaka, K. Ogura, S. Orimo, M. Kado, Y. Hayashi, M. Mori, S.V. Bulanov, T. Esirkepov, K. Nemoto, Y. Oishi, T. Nayuki, T. Fujii, A. Noda, Y. Iwashita, T. Shirai, and S. Nakamura. *The laser proton acceleration in the strong charge separation regime*. Physics Letters A **357**, 339 (2006).

[278] R.T. Eagleton, E.L. Clark, H.M. Davies, R.D. Edwards, S. Gales, M.T. Girling, D.J. Hoarty, N.W. Hopps, S.F. James, M.F. Kopec, J.R. Nolan, and K. Ryder. *Target diagnostics for commissioning the AWE HELEN Laser Facility 100 TW chirped pulse amplification beam*. Rev. Sci. Instrum. **77**, 10F522 (2006).

[279] M. Borghesi, J. Fuchs, S.V. Bulanov, A.J. Mackinnon, P.K. Patel, and M. Roth. *Fast ion generation by high-intensity laser irradiation of solid targets and applications*. Fusion Science and Technology **49**, 412 – 439 (2006).

[280] S. Humphries Jr. *Charged Particle Beams*. John Wiley and Sons., digital edition edition, (1990). Available online at http://www.fieldp.com/cpb.html (2008).

[281] R. Kodama, Y. Sentoku, Z.L. Chen, G.R. Kumar, S.P. Hatchett, Y. Toyama, T.E. Cowan, R.R. Freeman, J. Fuchs, Y. Izawa, M.H. Key, Y. Kitagawa, K. Kondo, T. Matsuoka, H. Nakamura, M. Nakatsutsumi, P.A. Norreys, T. Norimatsu, R.A. Snavely, R.B. Stephens, M. Tampo, K.A. Tanaka, and T. Yabuuchi. *Plasma devices to guide and collimate a high density of MeV electrons*. Nature **432**, 1005 (2004).

[282] NanoLabz, 661 Sierra Rose Dr., Reno, Nevada 89511, USA. Website, June (2008). Available online at http://www.nanolabz.com (2008).

[283] S. Gaillard, J. Fuchs, N. Renard-Le Galloudec, and T.E. Cowan. *Study of saturation of CR39 nuclear track detectors at high ion fluence and of associated artifact patterns*. Rev. Sci. Instrum. **78**, 013304 (2007).

[284] M. Roth, T.E. Cowan, J.C. Gauthier, J. Meyer ter Vehn, M. Allen, P. Audebert, A. Blazevic, J. Fuchs, M. Geißel, M. Hegelich, S. Karsch, and A. Pukhov T. Schlegel. *Relativistic Laser Plasmas Generating Intense, Collimated Ion Beams*. GSI-Report **GSI-2001-1** (2001).

[285] T. Eichner, F. Grüner, S. Becker, M. Fuchs, D. Habs, R. Weingartner, U. Schramm, H. Backe, P. Kunz, and W. Lauth. *Miniature magnetic devices for laser-based, table-top free-electron lasers*. Physical Review Special Topics - Accelerators and Beams **10**, 082401 (2007).

[286] V. Yanovsky, V. Chvykov, G. Kalinchenko, P. Rousseau, T. Planchon, T. Matsuoka, A. Maksimchuk, J. Nees, G. Cheriaux, G. Mourou, and K. Krushelnick. *Ultra-high intensity- 300-TW laser at 0.1 Hz repetition rate.* Optics Express **16**, 2109 (2008).

[287] Computer Simulation Technology (CST) GmbH, Bad Nauheimer Str. 19, 64289 Darmstadt, Germany. *CST Studio Suite.* Website, (2008). Available online at http://www.cst.com (2008).

[288] Pulsar Physics, Burghstraat 47, 5614 BC Eindhoven, The Netherlands. *The General Particle Tracer code.* Website, (2008). Available online at http://www.pulsar.nl (2008).

[289] D.R. Welch, D.V. Rose, M.E. Cuneo, R.B. Campbell, and T.A Mehlhorn. *Integrated simulation of the generation and transport of proton beams from laser-target interaction.* Phys. Plasmas **13**, 063105 (2006).

[290] E. Brambrink, M. Roth, A. Blazevic, and T. Schlegel. *Modeling of the electrostatic sheath shape on the rear target surface in short-pulse laser-driven proton acceleration.* Laser Part. Beams **24**, 163 – 8 (2007).

[291] R.G. Evans, E.L. Clark, R.T. Eagleton, A.M. Dunne, R.D. Edwards, W.J. Garbett, T.J. Goldsack, S. James, C.C. Smith, B.R. Thomas, R. Clarke, D.J. Neely, and S.J. Rose. *Rapid heating of solid density material by a petawatt laser.* Appl. Phys. Lett. **86**, 191505 (2005).

[292] R. Batra and M. Sehgal. *Range of electrons and positrons in matter.* Physical Review B **23**, 4448 – 4454 (1981).

[293] C.T. Zhou, M.Y. Yu, and X.T. He. *Density effect on proton acceleration from carbon-containing high-density thin foils irradiated by high-intensity laser pulses.* J. Appl. Phys. **101**, 3302 (2007).

[294] W. Theobald, K. Akli, R. Clarke, J.A. Delettrez, R.R. Freeman, S. Glenzer, J. Green, G. Gregori, R. Heathcote, N. Izumi, J.A. King, J.A. Koch, J. Kuba, K. Lancaster, A.J. Mackinnon, M. Key, C. Mileham, J. Myatt, D. Neely, P.A. Norreys, H.-S. Park, J. Pasley, P. Patel, S.P. Regan, H. Sawada, R. Shepherd, R. Snavely, R.B. Stephens, C. Stoeckl, M. Storm, B. Zhang, and T.C. Sangster. *Hot surface ionic line emission and cold K-inner shell emission from petawatt-laser-irradiated Cu foil targets.* Phys. Plasmas **13**, 043102 (2006).

[295] D.S. Dorozhkina and V.E. Semenov. *Exact solution of the problem of quasineutral expansion into vacuum of a localized collisionless plasma with cold ions.* JETP Lett. **67**, 573 – 578 (1998). translated from Pis'ma Zh. Éksp. Teor. Fiz. **67**(8), 543 - 547 (1998).

[296] D.S. Dorozhkina and V.E. Semenov. *Exact solution of Vlasov equations for quasineutral expansion of plasma bunch into vacuum.* Phys. Rev. Lett. **81**, 2691 – 2694 (1998).

[297] D.V. Griffiths and I.M. Smith. *Numerical methods for engineers: a programming approach.* CRC Press, Boca Raton, (1991). ISBN 0-8493-8610-1. See also the Wikipedia article at http://en.wikipedia.org/wiki/Midpoint˙method (2008).

[298] P. Antici, J. Fuchs, M. Borghesi, L. Gremillet, T. Grismayer, Y. Sentoku, E. d'Humières, C.A. Cecchetti, A. Mančić, A.C. Pipahl, T. Toncian, O. Willi, P. Mora, and P. Audebert. *Hot and Cold Electron Dynamics Following High-Intensity Laser Matter Interaction.* Phys. Rev. Lett. **101**, 105004 (2008).

[299] B. Dromey, S. Kar, M. Zepf, and P. Foster. *The plasma mirror - A subpicosecond optical switch for ultrahigh power lasers.* Rev. Sci. Instrum. **75**, 645 (2004).

[300] A. Lévy, T. Ceccotti, P. D'Oliveira, F. Réau, M. Perdrix, F. Quéré, P. Monot, M. Bougeard, H. Lagadec, P. Martin, J.-P. Geindre, and P. Audebert. *Double plasma mirror for ultrahigh temporal contrast ultraintense laser pulses.* Optics Letters **32**, 310 (2007).

[301] Baifei Shen, Yuelin Li, M.Y. Yu, and John Cary. *Bubble regime for ion acceleration in a laser-driven plasma.* Phys. Rev. E **76**, 055402(R) (2007).

[302] Ming-Ping Liu, Hai-Cheng Wu, Bai-Song Xie, Jie Liu, Hong-Yu Wang, and M.Y. Yu. *Energetic collimated ion bunch generation from an ultraintense laser interacting with thin concave targets.* Phys. Plasmas **15**, 063104 (2008).

[303] T. Okada, A.A. Andreev, Y. Mikado, and K. Okubo. *Energetic proton acceleration and bunch generation by ultraintense laser pulses on the surface of thin plasma targets.* Phys. Rev. E **74**, 026401 (2006).

[304] M.B.H. Breese, D.N. Jamieson, and B.L. Doyle. *The use of solenoid lenses in a two-stage nuclear microprobe probe-forming system.* Nucl. Instrum. and Meth. B **188**, 261 (2002).

[305] J. Fuchs, P. Antici, M. Fazi, A. Lombardi, M. Migliorati, L. Palumbo, and P. Audebert. *Postacceleration Of Laser-Generated High Energy Protons Through Conventional Accelerator Linacs.* volume 1024, 90–95. AIP conference proceedings: Laser-Driven Relativistic Plasmas Applied For Science, Industry, and Medicine: The 1st International Symposium, Kyoto (Japan), (2008).

[306] Virtual Institute to investigate in a collaborative effort the generation of intense Particle Beams by Ultra-intense Lasers "VIPBUL", (2005 - 2008). Virtual Institute VI-VH-144 funded by the Helmholtz Association. Available online at http://www.vipbul.de (2008).

[307] S. Nakamura, M. Ikegami, Y. Iwashita, T. Shirai, H. Tongu, H. Souda, H. Daido, M. Mori, M. Kado, A. Sagisaka, K. Ogura, M. Nishiuchi, S. Orimo, Y. Hayashi, A. Yogo, A.S. Pirozhkov, S.V. Bulanov, T. Esirkepov, A. Nagashima, T. Kimura, T. Tajima, T. Takeuchi, A. Fukumi, Z. Li, , and Akira Noda. *High-Quality Laser-Produced Proton Beam Realized by the Application of a Synchronous RF Electric Field.* Jpn. J. Appl. Phys. **46**, L717 – L720 (2007).

[308] R.W. Lee, S.J. Moon, H.K. Chung, W. Rozmus, H.A. Baldis, G. Gregori, R.C. Cauble, O.L. Landen, J.S. Wark, A. Ng, S.J. Rose, C.L. Lewis, D. Riley, J.-C. Gauthier, and P. Audebert. *Finite temperature dense matter studies on next-generation light sources.* J. Opt. Soc. Am. B **20**, 770 (2003).

[309] S.H. Glenzer, G. Gregori, R.W. Lee, F.J. Rogers, S.W. Pollaine, and O.L. Landen. *Demonstration of Spectrally Resolved X-Ray Scattering in Dense Plasmas.* Phys. Rev. Lett. **90**, 175002 (2003).

[310] S.H. Glenzer, O.L. Landen, P. Neumayer, R.W. Lee, K. Widmann, S.W. Pollaine, R.J. Wallace, G. Gregori, A. Höll, T. Bornath, R. Thiele, V. Schwarz, W.-D. Kraeft, and R. Redmer. *Observations of Plasmons in Warm Dense Matter.* Phys. Rev. Lett. **98**, 065002 (2007).

[311] N.A. Tahir, P. Spiller, A.R. Piriz, A. Shutov, I.V. Lomonosov, **M. Schollmeier**, A. Pelka, D.H.H. Hoffmann, and C. Deutsch. *Studies of high energy density states using isochoric heating of matter by intense ions beams: The HEDgeHOB Collaboration* – accepted for publication. Physica Scripta (2008).

[312] Walter F Henning. *The future GSI facility*. Nucl. Instrum. and Meth. B **214**, 211 (2004).

[313] The HEDgeHOB collaboration. *Studies on high energy density matter with intense heavy ion and laser beams at GSI*, January (2005). Available online at http://www.gsi.de/forschung/pp/TP-2005 e.pdf.

[314] O.L. Landen, S.H. Glenzer, M.J. Edwards, R.W. Lee, G.W. Collins, R.C. Cauble, W.W. Hsing, and B.A. Hammel. *Dense matter characterization by X-ray Thomson scattering*. J. Quant. Spectrosc. Radiat. Transfer **71**, 465 – 478 (2001).

[315] **M. Schollmeier**, G. Rodriguez Prieto, F.B. Rosmej, G. Schaumann, A. Blazevic, O.N. Rosmej, and M. Roth. *Investigation of laser-produced chlorine plasma radiation for non-monochromatic X-ray scattering experiments*. Laser Part. Beams **24**, 335–346 (2006).

[316] E. Brambrink, T. Schlegel, G. Malka, K.U. Amthor, M.M. Aléonard, G. Claverie, M. Gerbaux, F. Gobet, F. Hannachi, V. Méot, P. Morel, P. Nicolai, J.N. Scheurer, M. Tarisien, V. Tikhonchuk, and P. Audebert. *Direct evidence of strongly inhomogeneous energy deposition in target heating with laser-produced ion beams*. Phys. Rev. E **75**, 065401(R) (2007).

[317] I.V. Lomonosov A.V. Bushman and V.E. Fortov. *Equations of State for Materials at High Energy Density [in russian]*. Institute of Problems of Chemical Physics, Chernogolovka, Moskow, (1992).

[318] M. Bonitz, G. Bertsch, V.S. Filinov, and H. Ruhl. *Introduction to Computational Methods in Many Body Physics*. Cambridge University Press, Cambridge, (2004).

[319] H. Ruhl and A. Kemp. *Classical particle simulations with the PSC code*, (2006).

[320] D. O'flaherty and M. Goddard. *AMD Opteron™ processor benchmarking for clustered systems*. Technical report, Advanced Micro Devices, Inc., (2003).

[321] Lawrence Livermore National Laboratories. *Advanced Simulation and Computing: Blue-Gene/L*. Website, May (2008). Available online at https://asc.llnl.gov/computing resources/bluegenel/ (2008).

Publications

Controlled Transport and Focusing of Laser-Accelerated Protons with Miniature Magnetic Devices
M. Schollmeier, S. Becker, M. Geißel, K.A. Flippo, A. Blažević, S.A. Gaillard, D. C. Gautier, F. Grüner, K. Harres, M. Kimmel, F. Nürnberg, P. Rambo, U. Schramm, J. Schreiber, J. Schütrumpf, J. Schwarz, N.A. Tahir, B. Atherton, D. Habs, B.M. Hegelich and M. Roth
Physical Review Letters **101**, 055004 (2008)

Laser beam-profile impression and target thickness impact on laser-accelerated protons
M. Schollmeier, K. Harres, F. Nürnberg, A. Blažević, P. Audebert, E. Brambrink, J.C. Férnandez, K.A. Flippo, D.C. Gautier, M. Geißel, B.M. Hegelich, J. Schreiber, and M. Roth
Physics of Plasmas **15**, 053101 (2008)

Laser ion acceleration with micro-grooved targets
Marius Schollmeier, M. Roth, A. Blazevic, E. Brambrink, J.A. Cobble, J.C. Fernandez, K.A. Flippo, D.C. Gautier, D. Habs, K. Harres, B.M. Hegelich, T. Heßling, D.H.H. Hoffmann, S. Letzring, F. Nürnberg, G. Schaumann, J. Schreiber, K. Witte
Nuclear Instruments and Methods in Physics Research A **577**, 186 - 190 (2007)

Plasma physics experiments at GSI
M. Schollmeier, M. Roth, G. Schaumann, A. Blazevic, K. Flippo, A. Frank, J.C. Fernandez, D.C. Gautier, K. Harres, T. Heßling, B.M. Hegelich, F. Nürnberg, A. Pelka, H. Ruhl, J. Schreiber, D. Schumacher, K. Witte, B. Zielbauer, and D.H.H. Hoffmann
Journal of Physics: Conference Series **112**, 042068 (2008)

Investigation of laser-produced chlorine plasma radiation for non-monochromatic X-ray scattering experiments
M. Schollmeier, G. Rodríguez Prieto, F.B. Rosmej, G. Schaumann, A. Blazevic, O.N. Rosmej and M. Roth
Laser and Particle Beams **24**, 335 - 346 (2006)

Probing warm dense lithium by inelastic x-ray scattering
E. García Saiz, G. Gregori, D.O. Gericke, J. Vorberger, B. Barbrel, R.J. Clarke, R.R. Freeman, S.H. Glenzer, F.Y. Khattak, M. Koenig, O.L. Landen, D. Neely, P. Neumayer, M.M. Notley, A. Pelka, D. Price, M. Roth, **M. Schollmeier**, C. Spindloe, R.L. Weber, L. van Woerkom, K. Wünsch and D. Riley
Nature Physics **4**, 940 - 944 (2008)

Studies of high energy density states using isochoric heating of matter by intense heavy ion beams: The HEDgeHOB Collaboration
N.A. Tahir, P. Spiller, A.R. Piriz, A. Shutov, I.V. Lomonosov, **M. Schollmeier**, A. Pelka, D.H.H. Hoffmann, C. Deutsch
Physica Scripta **T132**, 014023 (2008)

Development and calibration of a Thomson parabola with microchannel plate for the detection of laser-accelerated MeV ions
K. Harres, **M. Schollmeier**, E. Brambrink, P. Audebert, A. Blažević, K. Flippo, M. Geißel, D.C. Gautier,
B.M. Hegelich, F. Nürnberg, J. Schreiber, H. Wahl and M. Roth
Review of Scientific Instruments **79**, 093306 (2008)

Increased efficiency of short-pulse laser-generated proton beams from novel flat-top cone targets
K.A. Flippo, E. d'Humières, S.A. Gaillard, J. Rassuchine, D.C. Gautier, **M. Schollmeier**, F. Nürnberg, J.L. Kline, J. Adams, B. Albright, M. Bakeman, K. Harres, R.P. Johnson, G. Korgan, S. Letzring, S. Malekos, N. Renard-LeGalloudec, Y. Sentoku, T. Shimada, M. Roth, T.E. Cowan, J.C. Fernández and B. M. Hegelich
Physics of Plasmas **15**, 056709 (2008)

Laser-driven ion accelerators: Spectral control, monoenergetic ions and new acceleration mechanisms
K. Flippo, B.M. Hegelich, B.J. Albright, L. Yin, D.C. Gautier, S. Letzring, **M. Schollmeier**, J. Schreiber, R. Schulze and J.C. Fernández
Laser and Particle Beams **25**, 3 - 8 (2007)

Laser accelerated heavy particles - Tailoring of ion beams on a nano-scale
M. Roth, P. Audebert, A. Blazevic, E. Brambrink, J. Cobble, T.E. Cowan, J. Fernandez, J. Fuchs, M. Geissel, M. Hegelich, S. Karsch, H. Ruhl, **M. Schollmeier** and R. Stephens
Optics Communications **264**, 519 - 524 (2006)

Laser-produced proton beams as a tool for equation-of-state studies of warm dense matter
Anna Tauschwitz, E. Brambrink, J.A. Maruhn, M. Roth, **M. Schollmeier**, T. Schlegel and Andreas Tauschwitz
High Energy Density Physics **2**, 16 - 20 (2006)

Ultrashort-laser-produced heavy ion generation via target ablation cleaning
K.A. Flippo, B.M. Hegelich, M.J. Schmitt, C.A. Meserole, G.L. Fisher, D.C. Gautier, J.A. Cobble, R. Johnson, S. Letzring, J. Schreiber, **M. Schollmeier** and J.C. Fernández
Journal de Physique IV **133**, 1117 - 1122 (2006)

Ablation cleaning techniques for high-power short-pulse laser-produced heavy ion targets
K.A. Flippo, B.M. Hegelich, M.J. Schmitt, D.C. Gauthier, C.A. Meserole, G.L. Fisher, J.A. Cobble, R.A. Johnson, S.A. Letzring, J.C. Fernandez, **M. Schollmeier** and J. Schreiber
Procedings of SPIE **6261**, 626121 (2006)

Laser accelerated ions in ICF research prospects and experiments
M. Roth, E. Brambrink, P. Audebert, M. Basko, A. Blazevic, R. Clarke, J. Cobble, T.E. Cowan, J. Fernandez, J. Fuchs, M. Hegelich, K. Ledingham, B.G. Logan, D. Neely, H. Ruhl and **M. Schollmeier**
Plasma Physics and Controlled Fusion **47**, B841 - B850 (2005)

High energy heavy ion jets emerging from laser plasma generated by long pulse laser beams from the NHELIX laser system at GSI

G. Schaumann, **M.S. Schollmeier**, G. Rodriguez-Prieto, A. Blazevic, E. Brambrink, M. Geissel, S. Korostiy, P. Pirzadeh, M. Roth, F.B. Rosmej, A.Ya. Faenov, T.A. Pikuz, K. Tsigutkin, Y. Maron, N.A. Tahir and D.H.H. Hoffmann

Laser and Particle Beams **23**, 503 - 512 (2005)

Acknowledgements

Mein erster Dank gilt Herrn Prof. Dr. Markus Roth für die Überlassung der Arbeit, sowie für die Möglichkeit, diese am GSI Helmholtzzentrum für Schwerionenforschung GmbH unter besten Bedingungen duchführen zu können. Ich danke ihm für sein großes Engagement, sowie die großzügige Unterstützung mit Rat, Tat, Reise- und Sachmitteln während der gesamten Arbeit, ohne die diese Arbeit nicht hätte realisiert werden können.

Sehr großer Dank gilt auch Herrn Prof. Dr. Dr. h.c./RUS Dieter H. H. Hoffmann für die Aufnahme in die Abteilung Plasmaphysik der GSI sowie die immerwährende Unterstützung. Ich danke ihm besonders für die Unterstützung meiner Arbeit durch die Entscheidung, einen Rechencluster für die PIC-Simulationen anzuschaffen.

Ein besonders großes Dankeschön gilt Herrn Prof. Dr. Hartmut Ruhl für die Überlassung seines PIC-Codes sowie der steten fachlichen Unterstützung, der vielen Treffen in Bochum und in Darmstadt und den fruchtbaren Diskussionen, die stets motivierend waren und einiges Licht ins Dunkel brachten. Ich danke ihm sehr für seine Vorarbeit beim CPT und dem Vertrauen, die weitere Arbeit am CPT gemeinsam durchzuführen.

Für die außergewöhnlich gute Zusammenarbeit möchte ich den "Ionboys" Dipl.-Phys. Knut Harres und Dipl.-Phys. Frank Nürnberg ganz besonders danken. Viele fruchtbare Diskussionen sowie ihr unermüdliches Engagement bei den Experimenten und deren Auswertung trugen wesentlich zum Gelingen dieser Arbeit bei.

Herrn Dr. Erik Brambrink danke ich für die Vorarbeiten auf dem Gebiet der Laserionenbeschleunigung, die einen einfachen Start ermöglichten, sowie für die vielen kritischen Diskussionen, der Unterstützung bei den Experimenten in LULI und am PHELIX und der großen Hilfe bei den Veröffentlichungen. Herrn Dr. Patrick Audebert sei hierbei für die Unterstützung und einen großartig laufenden 100 TW-Laser gedankt.

Den Herren Dr. Juan C. Fernández, Dr. Kirk A. Flippo, D. Cort Gautier sowie Dr. B. Manuel Hegelich danke ich sehr für meine Teilnahme an den Experimenten in Los Alamos sowie die gute Zusammenarbeit und Unterstützung darüber hinaus. Ich danke besonders Dr. B. Manuel Hegelich und Dr. Kirk A. Flippo für das Teilhaben an den Ergebnissen während meines zweiwöchigen, unfreiwilligen Hotelaufenthalts in White Rock.

Ein großes Dankeschön geht an Dr. Matthias Geißel sowie Dr. Briggs W. Atherton für die Einladung nach Sandia und den schönen und produktiven vier Wochen. Ein ebenso großer Dank geht an Dr. Patrick Rambo, Mark W. Kimmel und Dr. Jens Schwarz für einen nahezu perfekt laufenden Laser.

Des weiteren Danke ich Herrn Dipl.-Phys. Stefan Becker für die Charakterisierung und speziell für die Simulationen der PMQs sowie für die fruchtbaren Diskussionen bei der Experimentauswertung. Hierbei danke ich auch den Herren Prof. Dr. Dieter Habs, Dr. Florian Grüner und ganz besonders Dr. B. Manuel Hegelich und Dr. Kirk A. Flippo für die Bereitstellung der PMQs für das Experiment in Sandia.

Den Herren Dr. Jörg Schreiber, Dr. Bernhard Zielbauer sowie Herrn Prof. Dr. Klaus Witte und dem PHELIX-Team danke ich für die spannende Strahlzeit, die letzten Endes dann doch brauchbare Ergebnisse geliefert hat. Herrn Dr. Jörg Schreiber danke ich des Weiteren für die Strahlzeit in LULI.

Den Herren Dr. Siegfried Glenzer und Dr. Paul Neumayer danke ich sehr für die Einladung ans Lawrence Livermore National Laboratory und die spannenden vier Wochen, die (leider) tiefe Einblicke in die Tücken bei Experimenten zur Röntgenthomsonstreuung lieferten.

Herrn Dr. Roland Repnow und dem Team vom Tandembeschleuniger am Max-Planck-Institut für Kernphysik in Heidelberg danke ich sehr für die unbürokratische, meist doch sehr kurzfristige Möglichkeit der RCF-Kalibrationsmessungen sowie der optimalen Hilfe bei selbigen.

Den Herren Dr. Stefan Haller, Dr. Thomas Roth und Dr. Walter Schön danke ich für die tatkräftige und kompetente Unterstützung beim Aufbau und Betrieb des Clusters an der GSI. Ebenso möchte ich mich bei Herrn Prof. Dr. Robert Roth für die Rechenzeit auf der "tnpfarm" herzlich bedanken. In diesem Zusammenhang geht ein großes Dankeschön an das "PP-IT-Kompetenzzentrum", den Herren Dipl.-Phys. Thomas Heßling und Dipl.-Phys. Alexander Hug, für die Hilfe bei Hard- und Softwarefragen jeglicher Art.

Ebenso danke ich allen Mitarbeitern der Abteilung Plasmapysik für die gemeinsame Arbeit und schöne Zeit. Für ihre konkrete Unterstützung danke ich besonders Dr. Abel Blažević, Dr. Gabriel Schaumann, Dr. Naeem A. Tahir, Dr. Olga N. Rosmej, Dipl.-Phys. Marc Günther, Dipl.-Phys. Alexander Pelka, Dipl.-Phys. Jörg Schütrumpf, Dipl.-Phys. Anke Otten, Dipl.-Phys. Alexander Schökel, Ina Alber, Dominik Kraus und Daniel Löb und allen anderen, die ich hier vergessen habe.

Zu guter Letzt bedanke ich mich bei meinen Eltern, die mir mein Studium ermöglichten und mich auf vielfältige Art unterstützt und motiviert haben. Ein ganz großes Dankeschön geht an meine Frau Sabrina für die Liebe, die Geduld und das Verständnis, das sie mir in den Jahren entgegengebracht hat.

Die VDM Verlagsservicegesellschaft sucht für wissenschaftliche Verlage abgeschlossene und herausragende

Dissertationen, Habilitationen, Diplomarbeiten, Master Theses, Magisterarbeiten usw.

für die kostenlose Publikation als Fachbuch.

Sie verfügen über eine Arbeit, die hohen inhaltlichen und formalen Ansprüchen genügt, und haben Interesse an einer honorarvergüteten Publikation?

Dann senden Sie bitte erste Informationen über sich und Ihre Arbeit per Email an *info@vdm-vsg.de*.

Sie erhalten kurzfristig unser Feedback!

VDM Verlagsservicegesellschaft mbH
Dudweiler Landstr. 99   Telefon +49 681 3720 174
D - 66123 Saarbrücken   Fax   +49 681 3720 1749
www.vdm-vsg.de

Die VDM Verlagsservicegesellschaft mbH vertritt

Printed by Books on Demand GmbH, Norderstedt / Germany